Extreme Science
Transplanting Your Head

Extreme Science

Transplanting Your Head
and Other Feats of the Future

The Editors of SCIENTIFIC AMERICAN

Peter Jedicke, Project Editor

ST. MARTIN'S GRIFFIN

NEW YORK

A Byron Preiss Book

A Byron Preiss Book.
Editor: Peter Jedicke
Editorial Project Coordinator: Howard Zimmerman
Assistant Editor: Carlos Hiraldo
Design by Tom Draper Design
www.stmartins.com

The book's essays first appeared in the pages of *Scientific American Magazine* as follows:
"Making Methuselah," Fall 1999; "Growing New Organs," Fall 1999; "Tissue
Engineering," April 1999; "How Cancer Arises," September 1999; "Outbreak Not
Contained," April 2000; "Invasion of the Body Snatchers," May 1998; "Deadly Enigma,"
December 1996; "By the Numbers: Health Care Costs," April 1999; "Mutations Galore,"
April 1999; "Embryonic Stem Cells for Medicine," SCIAM Presents, Fall 1999;
"Encapsulated Cells as Therapy," April 1999; "Couture Cures: This Drug's for You,"
SCIAM Presents, Fall 1999; "Personal Pills" October 1998; "The Human Genome
Business Today," July 2000; "The Other Genomes," July 2000; "Designer Genomes,"
SCIAM Presents, Winter 1999; "Gene Therapy," October 1996; "Overcoming the
Obstacles to Gene Therapy," June 1997; "Gene Therapy for Cancer," June 1997; "Gene
Therapy for AIDS," June 1997; "I, Clone," SCIAM Presents, Fall 1999; "Cloning for
Medicine," December 1998; "Is Quiescence the Key to Cloning?" December 1998;
"Biochemist Baruch S. Blumberg: The Search For Extreme Life," July 2000; "Car Parts
from Chickens," April 2000; "Building a Brainier Mouse," April 2000; "The Basics: A
Mouse Named *Doogie*," April 2000; "Testing *Doogie*: Putting the Smart Mouse Through
Its Paces," April 2000; "The Search for a Memory-Boosting Drug," April 2000;
"Combinatorial Chemistry and New Drugs," April 1997; "Building the Better Bug,"
December 1998; "Head Transplants," SCIAM Presents, Fall 1999;
"The Coming Merging of Mind and Machine," SCIAM Presents, Fall 1999; "Nosing
Out a Mate," SCIAM Presents, Fall 1999; "Are You Ready for a New Sensation?" SCIAM
Presents, Fall 1999.

Library of Congress Cataloging-in-Publication Data

Extreme Science: Transplanting your head and other feats of the future / the editors of
Scientific American; Peter Jedicke, project editor.
    p.   cm.
    Includes bibliographical references and index.
    ISBN 0-312-26819-X
1. Medicine—Forecasting. 2. Biotechnology—Forecasting. I. Jedicke, Peter. II. Series.

R855.3 .T73 2001
610'.1'12—dc21

2001042405

10 9 8 7 6 5 4 3 2 1

First Edition: November 2001

CONTENTS

Chapter 8: Taking Humankind to the Extreme

Preface

Science touches every aspect of our lives, but there is enduring concern about the application of pure knowledge to practical technology. As this transfer is accomplished ever more speedily, our world often struggles to absorb the importance of new knowledge before the impact of the developing technology is felt. This is a struggle that we need to win in the 21st century. Our best hope is to follow science as the feats of the future unfold. If we understand change, we can be prepared for change. The articles in this book, selected from the pages of *Scientific American*, make that point vividly.

It is when science probes life itself that the most frightening prospects occur to us. Discoveries about quarks or quasars seem just too distant from everyday life to be worrisome. After all, life is what it's about. When Matthias Schleiden and Theodor Schwann discovered in the late 1830s that living things, whether humans, germs or trees, were made of cells, that was a radical forward stride. There has been a century and a half to assimilate the notion that there is a basic structure of life. Perhaps we are ready to use stem cells in medicine, or bacteria to prevent corrosion in pipes (see Chapter 3). Then just a few decades ago, James D. Watson and Francis Crick unraveled the DNA molecule, and that turned scientists loose on cloning and on the human genome. So here we are, thinking about genetically modified food (see Chapter 2), cloned sheep (see Chapter 8), and even the entire human genome (see Chapter 4). Gene therapy (see Chapter 5) may soon offer us the chance to fix certain problems that arise from faulty bits right in our own hereditary material. Are we ready for all this?

Although it all seems so new, the threat of change has been faced many, many times in the history of science. So we should know a little about Paul Ehrlich. Not the Paul Ehrlich who wrote about population in the 1960s and 1970s, but Paul Ehrlich, the father of wonder drugs, who shared the Nobel prize for medicine in 1908. Born in 1854, he started his medical studies with an interest in staining bacteria so that they could be seen more easily through a microscope. A natural talent for laboratory work landed him a professorship at the University of Berlin and even the directorship of an institute. In the 1890s, Ehrlich pursued an extreme idea: if chemical dyes could stain bacteria, perhaps a chemical related to the dyes could enter the same way as the dyes—and kill the nasty microbes. He called such chemicals "magic bullets" because they aimed themselves.

The theory was very successful. Ehrlich coined the term *chemotherapy* to describe this technique. Those were the days before IPOs, but Ehrlich invested a lot of money from the government and from a pharmaceutical company into a search for suitable chemicals that would attack various agents that caused disease. (It ended up making him a very wealthy man, regardless, and it is not entirely irrelevant to note that his widow had much of her wealth taken away by the Nazi regime. You see, Ehrlich was a Jew.) Arsenic being a famous poison, Ehrlich and his research assistants tested hundreds of compounds, old and new, that contained arsenic atoms. They were looking for molecules that would zero in on bacteria but leave healthy cells in the same organism alone.

You never know which way a story like this will turn. More than one disease like sleeping sickness was cured by Ehrlich's techniques. The number 606 is famous in bacteriology even today, because Ehrlich's team numbered the compounds they tested, and they discovered that compound 606 was most efficacious against spirochetes. Spirochetes are the organisms that cause syphilis, as Fritz Schaudinn found in 1905, among other diseases.

The development of compound 606 was a classic case of scientific progress: a clever hunch led to a long list of new chemicals, each of which was carefully tested. Thorough records were kept—all the more interesting today as historical documents. Over a few weeks in 1909, Ehrlich's lab injected compound 606 into rabbits that had been infected with syphilis, and the rabbits were cured. A few months later, Ehrlich passed compound 606 on to an associate who supervised a lunatic asylum, and patients whose syphilis cases were considered "hopeless" were given the chemical. Some of the patients responded quickly, and the promise of compound 606 was immediately recognized. Compound 606 was given the name salvarsan and was released to the medical world in December 1910.

The transition to proven technology was not without challenge. The chemical process that produced salvarsan was complex, so quality control was critical. Problems with the side effects of any drug based on arsenic made it imperative to use salvarsan cautiously, and one of Ehrlich's greatest concerns was the calculation of an appropriate dose in each case. Salvarsan oxidized quickly in air, and the oxidation products were deadly, so it had to be kept fresh and used right away. Perhaps most important of all, many doctors did not appreciate that it was necessary to use distilled water to prepare an injection of salvarsan, so bacteria in the water caused problems that had nothing to do with salvarsan's effectiveness.

What do you think was the reaction at the time to this significant breakthrough? Yes, many doctors wanted to apply salvarsan to sick patients right away. But imagine something like AIDS in a Victorian culture that only ever mentioned sex in whispers; that's the way it was with syphilis a century ago. For Ehrlich and the medical community to promote a cure for syphilis meant first raising awareness of the disease. For not the first time and certainly not the last, progress in science faced a formidable social barrier. Some denounced salvarsan because they favored "natural" remedies, some simply because they misunderstood its effectiveness. There were accusations of profiteering by the company that manufactured salvarsan. As it turned out, the controversy over salvarsan helped make discussion of sexually transmitted disease acceptable. Although the effectiveness of salvarsan was eventually superceded by that of penicillin, salvarsan was produced until 1974, and perhaps millions of patients were helped.

Can we explain why there was resistance to the introduction of salvarsan? More broadly, what scares non-scientists about science? Perhaps there is an even greater fear than the fear of the unknown: the fear of what we'll do to ourselves with what we learn. Insecure Victorians must have nibbled their cuticles raw, worrying about the collapse of morality that would result from a cure for syphilis. And the insecurity extended far beyond sexual lifestyles. The march of science has threatened morality, nationalism, pacifism, and more. Early in the 19th century, Mary Shelley expressed it better than anyone in *Frankenstein*—a tale of science turned monstrously against us. Mary Shelley had read the *Treatise on Sensations* by Etienne Bonnot, Abbé de Condillac, a book that appeared in 1754. In the book, de Condillac described how the thoughts that a statue could have, if the inhuman statue were endowed with human senses, might permit the statue to develop a human mind. Victor Frankenstein's creature was like that statue, but its horrible life brought only suffering and misery. Thank goodness we are not resigned to such a bleak and pessimistic view of technology, or no one would bother to make discoveries like salvarsan.

It is amazing how little things have changed—considering how much things have changed. The specter of biological warfare (see Chapter 3) is a very real monster. Philosophers still don't have a consensus on whether de Condillac's statue could develop a mind, but here we are, considering the possibility that a mind and a computer could be merged (see Chapter 8). Perhaps extending the human lifespan will bring strange new frictions (see Chapter 1). And what would happen if we were to lose control of our food supply (see Chapter 2)?

The articles presented here have been selected because these are the topics that challenge our world at the onset of the 21st century. Certainly we face risks, but science is a noble enlightenment, and the opportunities that lie before us are rich and rewarding. Whether it's a high-tech check-up at the doctor's office, a smart house or car parts made from chickens, here comes the future. Let's be ready.

—*Peter Jedicke*

Introduction:
Your Bionic Future

Television and slot machines notwithstanding, the point of technology is to extend what we can do with our bodies, our senses and, most of all, our minds. In the century just ended, we have gone from gaping at electric light-bulbs and telephones to channel-surfing past images of a sunrise on Mars, to outbursts of pique if our e-mail takes more than a few minutes to get to the other side of the world.

And in the next decade or two, the revolution is finally going to get really interesting. Several of the most important but disparate scientific and engineering achievements of the 20th century—the blossoming of electronics, the discovery of DNA and the elucidation of human genetics—will be the basis for leaps in technology that will extend, enhance or augment human capabilities far more directly, personally and powerfully than ever before.

The heady assortment of biotechnologies, implants, wearables, artificial environments, synthetic sensations and even demographic and societal shifts defies any attempt at concise categorization. But as our title boldly proclaims, we couldn't resist resurrecting the word *bionics*, lately in a state of anachronistic limbo alongside the 1970s television adventures that made it a household word. *Bionics* often refers to the replacement of living parts with cybernetic ones, but more broadly it also means engineering better artificial systems through biological principles. That merger of the biological with the microelectronic is at the heart of most of the coming advances.

As scientists and engineers fully unleash the power of the gene and of the electron, they will transform bits and pieces of the most fundamental facets of our lives, including eating and reproducing, staying healthy, being entertained and recovering from serious illness. Big changes could even be in store for what we wear, how we attract mates and how we stave off the debilitating effects of getting older. Within a decade, we will see:

- A cloned human being. It is possible, in fact, that experiments are already under way in secret.
- An artificial womb for women who can't become—or don't want to be—pregnant.
- Replacement hearts and livers, custom-grown from the recipient's own versatile stem cells.
- Virtual reality that becomes far more vivid and compelling by adding the senses of smell and touch to those of sight and sound.
- Custom clothing, assembled automatically from highly detailed scans of the purchaser's body and sold at a cost not much higher than off-the-rack.
- Foods that counteract various ailments, such as noninsulin-dependent diabetes, cholera, high cholesterol or hepatitis B.
- A genetic vaccine that endows the user with bigger, harder muscles, without any need to break a sweat at the gym.

With only a few exceptions, the articles collected here extrapolate conservatively into the near future. Essentially all the predicted developments will follow directly from technologies or advances that have already been achieved in the laboratory. Take that genetic muscle vaccine. A University of Pennsylvania researcher has already exercised buff laboratory mice whose unnaturally muscular hind legs were created by injection. He has little doubt about the suitability of the treatment for humans.

The two exceptions to the mostly restrained tone are the articles by neurosurgeon Robert J. White ("Head Transplants") and engineer-entrepreneur Ray Kurzweil ("The Coming Merging of Mind and Machine"), each of whom stakes out a position that is controversial among his peers. White raises the possibility of making the Frankenstein myth a reality as he declares that medical science is now capable of transplanting a human head onto a different body. Kurzweil argues not only that machines will eventually have human thoughts, emotions and consciousness but that their ability to share knowledge instantaneously will inexorably push them far past us in every category of endeavor, mental and otherwise.

Regardless of whether we ever see Frankenstein's monster, much less conscious machines, we already have enough details of the more immediate bionic future to let us raise some of the deeper questions about what it means. Depending on your viewpoint, there are plenty of uncomfortable if not alarming possible outcomes. Athletic competition, for example, could devolve into baroque spectacles that decide, basically, whose genetic enhancements (and work ethic) are best. Of course, it would be difficult to argue that such games would be intrinsically less interesting than today's contests, which pretty much decide whose natural genes (and work ethic) are best.

Since the 1970s such possibilities have tended to inspire relatively dark cultural movements. Examples include an entire subgenre of dystopian science fiction and one mad bomber. Historians and philosophers, too, are more likely now to analyze the negative ramifications of technology or even to attribute the endeavor to odd or unwholesome urges. Perhaps no one has written more entertainingly on the subject than the scholar William Irwin Thompson. In his 1991 book *The American Replacement of Nature*, he wrote:

> In truth, America is extremely uncomfortable with nature; hence its culturally sophisticated preference for the fake and nonnatural, from Cheez Whiz sprayed out of an aerosol can onto a Styrofoam potatoed chip, to Cool Whip smoothing out the absence of taste in those attractively red, genetically engineered monster strawberries. Any peasant with a dumb cow can make whipped cream, but it takes a chemical factory to make Cool Whip. It is the technological process and not the natural product that is important, and if it tastes bad, well, that's beside the point, for what that point is aimed at, is the escape from nature.

In the next decade or two the flight from nature will soar to new heights. The bright side of this transformation is potentially dazzling enough to drown out some of the dark visions. That is always the hope, of course. But the case now is unusually strong even if we base it on nothing more than the likelihood of powerful, sophisticated treatments for a host of dread genetic diseases and the frailties of old age. Those willing to grasp the implications of the coming fusion of biology and technology, with all its potential for beneficence and havoc, will find the exercise exhilarating.

—*Glenn Zorpette and Carol Ezzell, editors,* Scientific American

Chapter 1

Our Bodies in the Future

Making Methuselah

Karen Hopkin

"Most people are interested in living long and fruitful lives," begins the TV talk-show host, glancing at his notes.

"Fruit is good," interrupts the 2000-Year-Old Man. "Fruit kept me going for one hundred forty years once when I was on a very strict diet. Mainly nectarines. I love that fruit. Half a peach, half a plum. It's a hell of a fruit."

In their classic 1950s comedy routine, Carl Reiner and Mel Brooks had at least part of it figured out: we all want to live long and fruitful lives. But the answer may not lie in nectarines.

It may lie in worms. Or, more specifically, in what scientists are learning about longevity as they study organisms as diverse as roundworms, fruit flies, monkeys and humans. Their findings lend hope to those who think we might someday be able to slow the process of human aging.

"We can markedly increase the life span of simple organisms," reports Judith Campisi of Lawrence Berkeley National Laboratory. Researchers have found mutant worms, for example, that live up to 20 weeks—that's about eight times their normal life span and the equivalent of 600 years for you and me. They have also discovered treatments that can make normal human or animal cells grown in dishes live forever. And they have developed diet regimens that can increase life span while making animals healthier (though not necessarily happier).

"We're undergoing a major scientific revolution in our understanding of aging," maintains Michael R. Rose of the University of California at Irvine. But will any of these developments translate into a sip from the fountain of youth? Will scientists ever come up with a simple pill that will keep you looking good and feeling fine into the triple digits? Or—gasp!—even forever?

Questions such as these capture the imagination—and spark heated debate. "Our studies suggest that the rate at which animals age is not fixed in stone or immutable," states Cynthia Kenyon of the University of California at San Francisco. Kenyon has identified mutations that vastly increase the life span of roundworms. "By changing a few genes," she continues, "we can outwit death and keep the worms alive and youthful much longer." Simply mutating genes that control the way these worms respond to hormones that resemble insulin, for instance, enables them to live two to five times longer. A treatment that produced similar results might work for people, too. "If we can make it to ninety," she surmises, "I see no reason why, in principle, we couldn't make it to twice that."

Other scientists are less optimistic, though. "Such gene manipulations merely postpone the initiation of the aging process," declares U.C.S.F.'s Leonard Hayflick. "Aging is inevitable. Everything ages, including the universe." In 1961 Hayflick discovered that normal human cells, when grown in a culture dish, divide a limited number of times (about 50) and then die. This ultimate ceiling has been dubbed the Hayflick limit. "Saying that in twenty years we'll all live to be two hundred is utter nonsense," Hayflick says.

The Triumph of Entropy

First off, there's a difference between life span and life expectancy. Life expectancy, the number that appears on an insurance company actuarial table, reflects the average number of years a person can expect to live. Life span represents maximum longevity—the absolute number of years any human could hope to survive. The good news is that life expectancy has been on the rise for some time. People now live into their seventies, on average, which wasn't always the case. "99.99999 percent of the time humans have inhabited this planet, our life expectancy at birth has been no more than eighteen to twenty years," Hayflick notes. The increase we enjoy now is largely the result of humankind having conquered many infectious diseases. What is more, studies show that we're living not only longer but healthier, according to Richard J. Hodes, director of the National Institutes of Health's National Institute on Aging. As a population, we are less plagued than ever before by physical infirmity, muscle wasting, osteoporosis and the like.

But how old can we possibly live to be? Tests of simple animals such as Kenyon's worms suggest there may be no upper limit, observes Rose, who studies aging in flies.

"It's hard to imagine, though, that we could live past two hundred," says Leonard P. Guarente of the Massachusetts Institute of Technology, who has correlated a mutation that accelerates aging in yeast with a premature-aging syndrome in humans. "If we extend life span even a few years, cancer will kill everybody." And even if we duck cancer, he continues, wear and tear will weaken our veins and arteries, and our organs will eventually have to be patched up or replaced.

Even eliminating the diseases that now kill us would not increase our life expectancy substantially, Hayflick argues. Cure heart disease, add a dozen years; cancer, two or three more, he claims. "So if you cured both tomorrow morning, you'd only increase life expectancy by another fifteen years. That's it, period. End of sentence." Hayflick believes that the human life span may be fixed by our genes at an upper limit of about 125 years.

Our maximum life span may have become set during evolution, because there is really no need for any creature to live beyond its reproductive years. Humans escape this seemingly cruel contract, generally speaking, because we have no natural predators hunting down the infirm or elderly members of our society. As far as evolution is concerned, by the time an animal bears children, it has fulfilled its biological destiny to pass on its genes and is just taking up space and sponging off its kids.

In any case, evolutionarily speaking, there must be a price to be paid for longevity, suggests Steven Austad of the University of Idaho, who studies aging in wild mice, opossums and birds. "Otherwise we'd all be long-lived."

But maybe we only make that argument because we're one of the longest-lived animals around, Kenyon counters. "If we were dogs, we'd look at humans and think, 'Hey, they live for a really long time, why can't we?'"

Even if natural selection did not favor the evolution of humans with the longest life spans, Hodes declares, "there's no reason why we can't change that." But to come up with potential therapies to slow or halt aging, we first need to understand why we age.

Beginning at the End

By now almost everyone has heard of telomeres—the bits of repetitive DNA sequences that cap and protect the ends of our chromosomes. Even the border guard who checked Kenyon's passport as she crossed into Canada to attend a recent conference on aging emitted a knowing "Ah, telomeres" when she described the purpose of her visit. But how do telomeres relate to aging?

There's no doubt that telomeres are important for keeping cells alive in culture dishes in a laboratory. Allow connective tissue cells called fibroblasts to grow in culture and their telomeres get shorter and shorter each time the cells divide. And when a cell's telomeres shorten enough, they signal the cell to stop dividing. Activate telomerase—an enzyme that rebuilds telomeres—and cultured cells become immortal. Cancer cells can keep dividing in part because they reactivate their telomerase.

But is telomere shortening involved in aging in the body? It's debatable. In the body, telomeres do dwindle in size as cells age, eventually shrinking to a length that would signal the same cells to stop dividing in a culture dish. But there's no direct evidence that human cells stop growing in the body because their telomeres are too short, Guarente points out. "Cells from old people grow just fine in culture," he says. And as far as we know, Austad adds, "animals don't typically die because their cells don't divide any longer."

Still, researchers who earn their living studying telomeres are hedging their bets. "It's simply too early to judge," asserts Titia de Lange of the Rockefeller University. "We just do not know enough about telomeres and aging in humans."

That's where the mice come in. To examine more directly the link between telomeres and aging, Ronald A. DePinho of the Dana-Farber Cancer Institute in Boston has generated mice that lack telomerase and found that as these animals age their telomeres shrink. They also go gray and lose their hair—a result that de Lange deems "remarkable." The rodents do not, however, develop many of the other maladies generally considered hallmarks of aging, such as cataracts, osteoporosis and cardiac disease. DePinho's conclusion: "Telomere shortening is not the cause of overall aging as we know it."

But certain cells or tissues—especially those that are dividing rapidly— probably do become crippled by shortened telomeres, suggests Calvin L. Harley of Geron Corporation in Menlo Park, California. Withered telomeres might help weaken the immune system, bones or skin, for example, all of which contain rapidly dividing cells and all of which are compromised as we age. In these cells, telomere shrinkage may reach a critical point, after which chromosomes begin to break. So someday doctors might boost immune function or strengthen bone or skin by turning on telomerase in the appropriate cells. Telomerase might also help extend the lives of the rapidly dividing endothelial cells that line blood vessels, allowing them to repair the wear and tear caused by a lifetime of vigorous blood flow.

Having long, luxuriant telomeres also seems to help animals deal with stress, DePinho posits. In his telomerase-deficient mice, old age and telomere loss act together to reduce the animals' ability to handle and survive stress, such as chemotherapy. Dwindling telomeres, he concludes, might explain why older people tend to have trouble recovering from surgery, infections or wounds. In the future, DePinho foresees, perhaps cancer patients scheduled for chemotherapy will also receive telomerase to prevent the treatment's side effects and enable their blood cells to survive and proliferate.

But would switching on telomerase all over the body allow people to live to the ripe old age of 150? "I doubt it," Harley declares. "When it comes to maximum human life span, so many other factors could be involved."

Oxygen: A Deadly Gas

Take free radicals, for example. Scientists have hypothesized since the 1950s that destructive molecules called free radicals might contribute to aging. These molecules—which are generated as by-products of breaking down oxygen—can damage almost every critical component of cells, including DNA, proteins, and the fatty compounds that make up the inner and outer membranes of cells.

"Oxygen is toxic," declares Rajindar Sohal of Southern Methodist University. And the rate at which an animal ages may relate to how well it detoxifies oxygen radicals. Sohal finds that aged flies accumulate specific types of free-radical damage in their mitochondria—the tiny subcellular organelles that provide power to cells and tissues, including a fly's flight muscles. Martin Chalfie of Columbia University recently found that worms that lack a newly discovered form of an enzyme called catalase do not live as long as normal worms. Catalase disposes of hydrogen peroxide, a chemical that cells generate as they are converting oxygen into water. Further, Irvine's Rose has bred flies that live twice as long as normal. He finds that they show, among other things, an increase in the activity of superoxide dismutase (SOD)—an enzyme that destroys toxic oxygen radicals called superoxides.

Free radicals might also explain why pigeons live 35 years, 12 times longer than rats, animals that are about the same size. For the amount of oxygen they take in, pigeons produce fewer free radicals than rodents do. Perhaps we should be studying these animals to see how nature solves the aging problem, Austad suggests.

In the case of free radicals and aging, researchers need to be mindful of whether they are seeing cause and effect or simply a correlation, Guarente warns. Sure, oxygen radicals and cellular damage increase with age. But just because antioxidants increase life expectancy does not mean that free radicals cause aging. Banning motor vehicles would increase our life expectancy by about six months, Hayflick notes: "But that doesn't mean cars cause aging."

Free radicals can't be the bottom line when it comes to aging, Campisi agrees. "Mice and men live in the same toxic world."

So is SOD therapy likely in our future? "There's no guarantee it will work in humans," Rose admits. How about taking megadoses of antioxidants, such as vitamins C and E? That may not be good either, cautions Hodes, who recalls a study in which a group of smokers given the antioxidant beta carotene actually developed more cancers than a group of control subjects did.

No Sex + Less Food = Long Life

Arguably the most striking results of studies examining ways to boost longevity come from investigations of the simplest organisms. Kenyon, for example, looks at worms that live two, three or four times longer than average. The creatures' longevity seems to boil down to the way they respond to hormones similar to insulin. Somehow mutations in this pathway allow these worms to stay frisky and svelte way past their prime, explains Kenyon, who adds, "I don't think, at the molecular level, we have much idea how."

Interestingly, she finds that removing the animals' sperm and egg cells does the same thing. Mature sex cells accelerate aging, perhaps by producing the insulinlike hormones that seem to control longevity in worms, Kenyon observes. Such an arrangement may allow animals that mature slowly to remain healthy long enough to reproduce.

This dovetails nicely with what Rose finds in his flies. He breeds longer-lived flies by delaying when the insects reproduce. "Like 'good' teenagers, they don't waste their energy on sex," he reports. As a result, they have more verve left for later. When these flies are 40 or 50 days old—over the hill in human terms—"they're flying around, fornicating and having a good time while the regular flies are dying," Rose says.

Does that mean people should put off having kids? "Oh, no, that's totally impractical," Rose responds. "What I'm doing to these flies is much more severe than what career women are doing." Besides, delaying parenthood would not affect your own life span—although it might help your descendants live it up 100 generations down the line.

The caveat? Scientists need to be certain that they are not looking at interventions that merely decrease metabolic rate, which also increases life span. Put a fly in the fridge, and it will live eight or nine times as long, Sohal states. But humans probably would not want to live longer if they had to chill out and hibernate. Although Rose's flies appear to have the same metabolic rate as adults, DePinho insists, "we need to bring these findings back to mammalian systems to see how relevant they are."

So far the only intervention that has been proved to slow aging in mammals is calorie restriction. Mice and rats raised on a diet high in nutrition but reduced in calories by 30 to 60 percent live about 30 percent longer—and by all accounts are healthier to boot, reports Richard H. Weindruch of the University of Wisconsin. In addition to his work with rodents, Weindruch has been following a colony of rhesus monkeys that have been on a restricted diet for 10 years. Compared with nondieting animals, these middle-aged monkeys have low insulin levels and are better able to regulate their glucose. They also have lower triglyceride levels, which means they are probably less prone to developing atherosclerosis, another benefit that might allow them to live longer.

The food-restricted monkeys also have less free-radical damage to their skeletal muscles than animals that are allowed to eat their fill. Together, these results suggest that the researchers who are finding that insulin regulation and oxygen radicals are important in aging in flies and worms are on to something.

But calorie restriction won't necessarily lead to another new "miracle" diet. "Nobody proposes that we starve people so they live to be one hundred and fifty," Campisi counters. And the truth is that this diet would not be easy for people to pull off, Weindruch admits. It's tricky to cut that many calories and still maintain a nutritious diet. But if scientists can catalogue the physiological changes that occur in these animals, they may be able to design an intervention that accomplishes the same thing in humans who won't give up their Häagen-Dazs.

Pill Me

What does all this presage for potential antiaging therapies? The findings in calorie-restricted mammals suggest that to some degree longevity hinges on the hormones that control glucose metabolism, notes Richard A. Miller, a pathologist who studies aging mice at the University of Michigan School of Medicine. And the worm studies reveal that related hormonal pathways might regulate aging in all organisms. Animals that burn glucose more efficiently—extracting more energy from less blood sugar—somehow manage to live longer and healthier lives, Austad adds. This raises the possibility that therapies aimed at manipulating hormones might put the brakes on aging—or perhaps stave off aging-related ills such as osteoporosis, muscle loss, heart disease and cancer.

But even manipulating hormones may not be the whole answer. At the very least, we will need two different antiaging interventions, Guarente proposes: one for the brain and heart—cells that do not divide much—and another for cells that divide rapidly, such as skin. That is, unless you just want to look good. Adding telomerase might stretch the lives of skin cells, for example, but heart cells may need to be protected from the ravages of free radicals by somehow shoring up antioxidant defenses or regulating glucose metabolism.

"There's not going to be a magic bullet" to beat Father Time, Rose predicts. Campisi agrees. "To think that a single pill would slow all aging is extremely naive," she says. But someday certain interventions may be used to help particular systems of the body last longer and to prevent some age-related disorders. Retarding the death of neurons may not dramatically extend life span, for instance, but it might delay the onset of neurodegenerative diseases such as Alzheimer's disease so that they do not appear until age 90 or 100.

And as with anything, living longer may have its price. So-called dwarf mice, which are about one-third the size of normal mice and live 50 to 70 percent longer, are sterile. Calorie restriction delays puberty in rats, mice and monkeys. And the maggots produced by long-lived flies die in greater numbers than those of normal flies do. "So we're never going to see childhood immunization against aging," Austad advises. But therapy later in life, after childbearing, might be an option.

Just beware the quick fix, Miller warns. Most of the people who will tell you that we can prolong the human life span are "quacks who have something to sell." If Austad were less scrupulous, he might be among them. "I like the royal jelly idea," he comments. People eat this gooey substance because bees feed it to their queens and queens live longer than drones, he says. "But mostly it's just bee poop." Perhaps the fact that researchers who study aging aren't getting rich hawking antiaging therapies suggests that they haven't found the answers—yet.

"Right now aging is still very much a black box," Guarente admits. "But we're standing on the brink of understanding." Chalfie predicts that "we'll learn a staggering amount about the biology of aging in the next fifty years. What we'll be able to do with that information, it's hard to say."

Growing New Organs

David J. Mooney and Antonios G. Mikos

Every day thousands of people of all ages are admitted to hospitals because of the malfunction of some vital organ. Because of a dearth of transplantable organs, many of these people will die. In perhaps the most dramatic example, the American Heart Association reports that only 2,300 of the 40,000 Americans who needed a new heart in 1997 got one. Lifesaving livers and kidneys likewise are scarce, as is skin for burn victims and others with wounds that fail to heal. It can sometimes be easier to repair a damaged automobile than the vehicle's driver because the former may be rebuilt using spare parts, a luxury that human beings simply have not enjoyed.

An exciting new strategy, however, is poised to revolutionize the treatment of patients who need new vital structures: the creation of man-made tissues or organs, known as neo-organs. In one scenario, a tissue engineer injects or places a given molecule, such as a growth factor, into a wound or an organ that requires regeneration. These molecules cause the patient's own cells to migrate into the wound site, turn into the right type of cell and regenerate the tissue. In the second, and more ambitious, procedure, the patient receives cells—either his or her own or those of a donor—that have been harvested previously and incorporated into three-dimensional scaffolds of biodegradable polymers, such as those used to make dissolvable sutures. The entire structure of cells and scaffolding is transplanted into the wound site, where the cells replicate, reorganize and form new tissue. At the same time, the artificial polymers break down, leaving only a completely natural final product in the body—a neo-organ. The creation of neo-organs applies the basic knowledge gained in biology over the past few decades to the problems of tissue and organ reconstruction, just as advances in materials science make possible entirely new types of architectural design.

Science-fiction fans are often confronted with the concept of tissue engineering. Various television programs and movies have pictured individual organs or whole people (or aliens) growing from a few isolated cells in a vat of some powerful nutrient. Tissue engineering does not yet rival these fictional presentations, but a glimpse of the future has already arrived. The creation of tissue for medical use is already a fact, to a limited extent, in hospitals across the U.S. These groundbreaking applications involve fabricated skin, cartilage, bone, ligament and tendon and make musings of "off-the-shelf" whole organs seem less than far-fetched.

Indeed, evidence abounds that it is at least theoretically possible to engineer large, complex organs such as livers, kidneys, breasts, bladders and intestines, all of which include many different kinds of cells. The proof can be found in any expectant mother's womb, where a small group of undifferentiated cells finds the way to develop into a complex individual with multiple organs and tissues with vastly different properties and functions. Barring any unforeseen impediments, teasing out the details of the process by which a liver becomes a liver, or a lung a lung, will eventually allow researchers to replicate that process.

A Pinch of Protein

Cells behave in predictable ways when exposed to particular biochemical factors. In the simpler technique for growing new tissue, the engineer exposes a wound or damaged organ to factors that act as proponents of healing or regeneration. This concept is based on two key observations, in bones and in blood vessels.

In 1965 Marshall R. Urist of the University of California at Los Angeles demonstrated that new, bony tissue would form in animals that received implants of powdered bone. His observation led to the isolation of the specific proteins (the bone morphogenetic proteins, or BMPs) responsible for this activity and to the determination of the DNA sequences of the relevant genes. A number of biotechnology companies subsequently began to produce large quantities of recombinant human BMPs; the genes coding for BMPs were inserted into mammalian cell lines that then produced the proteins.

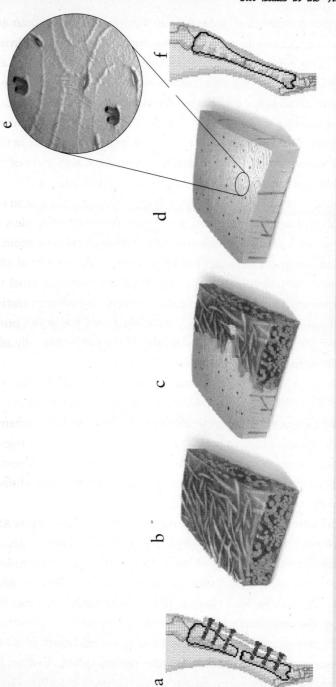

New bone grows to fill a space between two bone segments. A dog leg bone with a missing section is held in place with braces (a). A polymer scaffold primed with bone growth–promoting proteins (b) fills in the gap. The scaffold is slowly infiltrated by new bone (c) and ultimately gets completely replaced (d). The cells (e) have their own blood supply. After several months the leg bone has healed completely (f).

Various clinical trials are under way to test the ability of these bone growth promoters to regenerate bony tissue. Applications of this approach that are currently being tested include healing acute bone fractures caused by accidents and boosting the regeneration of diseased periodontal tissues. Creative BioMolecules in Hopkinton, Massachusetts, recently completed clinical trials showing that BMP-7 does indeed help heal severe bone fractures. This trial followed 122 patients with leg fractures in which the sections failed to rejoin after nine months. Patients whose healing was encouraged by BMP-7 did as well as those who received a surgical graft of bone harvested from another part of their body.

A critical challenge in engineering neo-organs is feeding each and every cell. Tissues more than a few millimeters thick require blood vessels to grow into them and supply the necessary nutrients. Fortunately, investigations by Judah Folkman have shown that cells already in the body can be coaxed into producing new blood vessels. Folkman, a cancer researcher at Harvard Medical School's Children's Hospital, recognized this possibility almost three decades ago in studies aimed, ironically, at the prevention of cellular growth in the form of cancerous tumors.

Folkman perceived that developing tumors need to grow their own blood vessels to supply themselves with nutrients. In 1972 he proposed that specific molecules could be used to inhibit such vessel growth, or angiogenesis, and perhaps starve tumors. (This avenue of attack against cancer became a major news story in 1998.) Realizing that other molecules would undoubtedly abet angiogenesis, he and others have subsequently identified a number of factors in each category.

That work is now being exploited by tissue engineers. Many angiogenesis-stimulating molecules are commercially available in recombinant form, and animal studies have shown that such molecules promote the growth of new blood vessels that bypass blockages in, for example, the coronary artery. Small-scale trials are also under way to test this approach in the treatment of similar conditions in human subjects.

Scientists must surmount a few obstacles, however, before drugs that promote tissue and organ formation become commonplace. To date, only the factors responsible for bone and blood vessel growth have been characterized. To regenerate other organs, such as a liver, for example, the specific molecules for their development must be identified and produced reliably.

An additional, practical issue is how best to administer the substances that would shape organ regeneration. Researchers must answer these questions: What specific concentrations of the molecules are needed for the desired effect? How long should the cells be exposed? How long will the factors be active in the body? Certainly multiple factors will be needed for complex organs, but when exactly in the development of the organ does one factor need to replace another? Controlled drug-delivery technology such as transdermal patches developed by the pharmaceutical industry will surely aid efforts to resolve these concerns.

In particular, injectable polymers may facilitate the delivery of bioactive molecules where they are needed, with minimal surgical intervention. Michael J. Yaszemski of the Mayo Clinic, Alan W. Yasko of the M.D. Anderson Cancer Center in Houston and one of us (Mikos) are developing new injectable biodegradable polymers for orthopedic applications. The polymers are moldable, so they can fill irregularly shaped defects, and they harden in 10 to 15 minutes to provide the reconstructed skeletal region with mechanical properties similar to those of the bone they replace. These polymers subsequently degrade in a controlled fashion, over a period of weeks to months, and newly grown bone fills the site.

We have also been studying the potential of injectable, biodegradable hydrogels—gelatinlike, water-filled polymers—for treating dental defects, such as poor bonding between teeth and the underlying bone, through guided bone regeneration. The hydrogels incorporate molecules that both modulate cellular function and induce bone formation; they provide a scaffold on which new bone can grow, and they minimize the formation of scar tissue within the regenerated region.

An intriguing variation of more conventional drug delivery has been pioneered by Jeffrey F. Bonadio, Steven A. Goldstein and their coworkers at the University of Michigan. (Bonadio is now at Selective Genetics in San Diego.) Their approach combines the concepts of gene therapy and tissue engineering. Instead of administering growth factors directly, they insert genes that encode those molecules. The genes are part of a plasmid, a circular piece of DNA constructed for this purpose. The surrounding cells take up the DNA and treat it as their own. They turn into tiny factories, churning out the factors coded for by the plasmid. Because the inserted DNA is free-floating, rather than incorporated into the cells' own DNA, it eventually degrades and the product ceases to be synthesized. Plasmid inserts have successfully promoted bone regrowth in animals; the duration of their effects is still being investigated.

One of us (Mooney), along with Lonnie D. Shea and our other aforementioned Michigan colleagues, recently demonstrated with animals that three-dimensional biodegradable polymers spiked with plasmids will release that DNA over extended periods and simultaneously serve as a scaffold for new tissue formation. The DNA finds its way into adjacent cells as they migrate into the polymer scaffold. The cells then express the desired proteins. This technique makes it possible to control tissue formation more precisely; physicians might one day be able to manage the dose and time course of molecule production by the cells that take up the DNA and deliver multiple genes at various times to promote tissue formation in discrete stages.

A Dash of Cells

Promoting tissue and organ development via growth factors is obviously a considerable step forward. But it pales in comparison to the ultimate goal of the tissue engineer: the creation from scratch of whole neo-organs. Science fiction's conception of prefabricated "spare parts" is slowly taking shape in the efforts to transplant cells directly to the body that will then develop into the proper bodily component. The best way to sprout organs and tissues is still to rely on the body's own biochemical wisdom; the appropriate cells are transferred, in a three-dimensional matrix, to the desired site, and growth unfolds within the person or organism rather than in an external, artificial environment. This approach, pioneered by Ioannis V. Yannas, Eugene Bell and Robert S. Langer of the Massachusetts Institute of Technology, Joseph P. Vacanti of Harvard Medical School and others in the 1970s and 1980s, is now actually in use in some patients, notably those with skin wounds or cartilage damage.

The usual procedure entails the multiplication of isolated cells in culture.

These cells are then used to seed a matrix, typically one consisting of synthetic polymers or collagen, the protein that forms the natural support scaffolding of most tissues. In addition to merely delivering the cells, the matrix both creates and maintains a space for the formation of the tissue and guides its structural development. Once the development rules for a given organ or tissue are known, any of those entities could theoretically be grown from a sample of starter cells. (A sufficient understanding of the development pathways should eventually allow the transfer of this procedure from the body to the laboratory, making true off-the-shelf organs possible. A surgeon could implant these immediately in an emergency situation—an appealing notion, because falling organs can quickly lead to death—instead of waiting weeks or months to grow a new organ in the laboratory or to use growth factors to induce the patient's own body to grow the tissues.)

In the case of skin, the future is here. The U.S. Food and Drug Administration has already approved a living skin product—and others are now in the regulatory pipeline. The need for skin is acute: every year 600,000 Americans suffer from diabetic ulcers, which are particularly difficult to heal; another 600,000 have skin removed to treat skin cancer; and between 10,000 and 15,000 undergo skin grafts to treat severe burns.

The next tissue to be widely used in humans will most likely be cartilage for orthopedic, craniofacial and urological applications. Currently available cartilage is insufficient for the half a million operations annually in the U.S. that repair damaged joints and for the additional 28,000 face and head reconstructive surgeries. Cartilage, which has low nutrient needs, does not require growth of new blood vessels—an advantage for its straightforward development as an engineered tissue.

Genzyme Tissue Repair in Cambridge, Massachusetts, has received FDA approval to engineer tissues derived from a patient's own cells for the repair of traumatic knee-cartilage damage. Its procedure involves growing the patient's cells, harvested whenever possible from the same knee under repair, in the lab, and then implanting those cells into the injury. Depending on the patient and the extent of the defect, full regeneration takes between 12 and 18 months. In animal studies, Charles A. Vacanti of the University of Massachusetts Medical School in Worcester, his brother, Joseph Vacanti, Langer and their colleagues have shown that new cartilage can be grown in the shapes of ears, noses and other recognizable forms.

The relative ease of growing cartilage has led Anthony J. Atala of Harvard Medical School's Children's Hospital to develop a novel approach for treating urological disorders such as incontinence. Reprogenesis in Cambridge, Massachusetts, which supports Atala's research, is testing whether cartilage cells can be removed from patients, multiplied in the laboratory and used to add bulk to the urethra or ureters to alleviate urinary incontinence in adults and bladder reflux in children. These conditions are often caused by a lack of muscle tone that allows urine to flow forward unexpectedly or, in the childhood syndrome, to back up. Currently patients with severe incontinence or bladder reflux may undergo various procedures, including complex surgery. Adults sometimes receive collagen that provides the same bulk as the cartilage implant, but collagen eventually degrades. The new approach involves minimally invasive surgery to deliver the cells and grow the new tissue.

Walter D. Holder, Jr., and Craig R. Halberstadt of Carolinas Medical Center in Charlotte, North Carolina, and one of us (Mooney) have begun to apply such general tissue-engineering concepts to a major women's health issue. We are attempting to use tissue from the legs or buttocks to grow new breast tissue, to replace that removed in mastectomies or lumpectomies. We propose to take a biopsy of the patient's tissue, isolate cells from this biopsy and multiply these cells outside the body. The woman's own cells would then be returned to her in a biodegradable polymer matrix. Back in the body, cell growth and the deterioration of the matrix would lead to the formation of completely new, natural tissue. This process would create only a soft-tissue mass, not the complex system of numerous cell types that makes up a true breast. Nevertheless, it could provide an alternative to current breast prostheses or implants.

Optimism for the growth of large neo-organs of one or more cell types has been fueled by success in animal models of human diseases. Mikos has demonstrated that new bone tissue can be grown by transplanting cells taken from bone marrow and growing them on biodegradable polymers. Transplantation of cells to skeletal defects makes it possible for cells to produce factors locally, offering a new means of delivery for growth-promoting drugs.

Recipes for the Future

In any system, size imposes new demands. As previously noted, tissues of any substantial size need a blood supply. To address that requirement, engineers may need to transplant the right cell types together with drugs that spur angiogenesis. Molecules that promote blood vessel growth could be included in the polymers used as transplant scaffolds. Alternatively, we and others have proposed that it may be possible to create a blood vessel network within an engineered organ prior to transplantation by incorporating cells that will become blood vessels within the scaffold matrix. Such engineered blood vessels would then need only to connect to surrounding vessels for the engineered tissue to develop a blood supply.

In collaboration with Peter J. Polverini of Michigan, Mooney has shown that transplanted blood vessel cells will indeed form such connections and that the new vessels are a blend of both implanted and host cells. But this technique might not work when transplanting engineered tissue into a site where blood vessels have been damaged by cancer therapy or trauma. In such situations, it may be necessary to propagate the tissue first at another site in the body where blood vessels can more readily grow into the new structure. Mikos collaborates with Michael J. Miller of the M.D. Anderson Cancer Center to fabricate vascularized bone for reconstructive surgery using this approach. A jawbone, for instance, could be grown connected to a well-vascularized hipbone for an oral cancer patient who has received radiation treatments around the mouth that damaged the blood supply to the jawbone.

On another front, engineered tissues typically use biomaterials that are available from nature or that can be adapted from other biomedical uses. We and others, however, are developing new biodegradable materials specific to this task. These may accurately determine the size and shape of an engineered tissue, precisely control the function of cells in contact with the material and degrade at rates that optimize tissue formation.

Structural tissues, such as skin, bone and cartilage, will most likely continue to dominate the first wave of success stories, thanks to their relative simplicity. The holy grail of tissue engineering, of course, remains complete internal organs. The liver, for example, performs many chemical reactions critical to life, and more than 30,000 people die every year because of liver failure. It has been recognized since at least the time of the ancient Greek legend of Prometheus that the liver has the unique potential to regenerate partially after injury, and tissue engineers are now trying to exploit this property of liver cells.

A number of investigators, including Joseph Vacanti and Achilles A. Demetriou of Cedars-Sinai Medical Center in Los Angeles, have demonstrated that new liverlike tissues can be created in animals from transplanted liver cells. We have developed new biomaterials for growing liverlike tissues and shown that delivering drugs to transplanted liver cells can increase their growth. The new tissues grown in all these studies can replace single chemical functions of the liver in animals, but the entire function of the organ has not yet been replicated.

H. David Humes of Michigan and Atala are using kidney cells to make neo-organs that possess the filtering capability of the kidney. In addition, recent animal studies by Joseph Vacanti's group have demonstrated that intestine can be grown—within the abdominal cavity—and then spliced into existing intestinal tissue. Human versions of these neointestines could be a boon to patients suffering from short-bowel syndrome, a condition caused by birth defects or trauma. This syndrome affects physical development because of digestion problems and insufficient nutrient intake. The only available treatment is an intestinal transplant, although few patients actually get one, again because of the extreme shortage of donated organs. Recently Atala has also demonstrated in animals that a complete bladder can be formed with this approach and used to replace the native bladder.

Even the heart is a target for regrowth. A group of scientists headed by Michael V. Sefton at the University of Toronto recently began an ambitious project to grow new hearts for the multitude of people who die from heart failure every year. It will very likely take scientists 10 to 20 years to learn how to grow an entire heart, but tissues such as heart valves and blood vessels may be available sooner. Indeed, several companies, including Advanced

Tissue Sciences in La Jolla, California, and Organogenesis in Canton, Massachusetts, are attempting to develop commercial processes for growing these tissues.

Prediction, especially in medicine, is fraught with peril. A safe way to prophesy the future of tissue engineering, however, may be to weigh how surprised workers in the field would be after being told of a particular hypothetical advance. Tell us that completely functional skin constructs will be available for most medical uses within five years, and we would consider that reasonable. Inform us that fully functional, implantable livers will be here in five years, and we would be quite incredulous. But tell us that this same liver will be here in, say, 30 years, and we might nod our heads in sanguine acceptance—it sounds possible. Ten millennia ago the development of agriculture freed humanity from a reliance on whatever sustenance nature was kind enough to provide. The development of tissue engineering should provide an analogous freedom from the limitations of the human body.

Tissue Engineering

Robert S. Langer and Joseph P. Vacanti

Tissue engineering has emerged as a thriving new field of medical science. Just a few years ago most scientists believed that human tissue could be replaced only with direct transplants from donors or with fully artificial parts made of plastic, metal and computer chips. Many thought that whole bioartificial organs—hybrids created from a combination of living cells and natural or artificial polymers—could never be built and that the shortage of human organs for transplantation could only be met by somehow using organs from animals.

Now, however, innovative and imaginative work in laboratories around the world is demonstrating that creation of biohybrid organs is entirely feasible. Biotechnology companies that develop tissue-engineered products have a market worth of nearly $4 billion, and they are spending 22.5 percent more every year. But before this investment will begin to pay off in terms of reliably relieving human suffering caused by defects in a wide range of tissues, tissue engineering must surmount some important hurdles.

Off-the-Shelf Cells

Establishing a reliable source of cells is a paramount priority for tissue engineers. Animal cells are a possibility, but ensuring that they are safe remains a concern, as does the high likelihood of their rejection by the immune system. For those reasons, human cells are favored.

The recent identification of human embryonic stem cells—cells that can give rise to a wide array of tissues that make up a person—offers one approach to the problem. But researchers are a long way from being able to manipulate embryonic stem cells in culture to produce fully differentiated cells that can be used to create or repair specific organs.

A more immediate goal would be to isolate so-called progenitor cells from tissues. Such progenitors have taken some of the steps toward becoming specialized, but because they are not yet fully differentiated they stay flexible enough to replenish several different cell types. Arnold I. Caplan of the Cleveland Clinic and his colleagues, for instance, have isolated progenitor cells from human bone marrow that can be prompted in the laboratory to form either the osteoblasts that make bone or the chondrocytes that compose cartilage. Similarly, Lola Reid of the University of North Carolina at Chapel Hill has identified small, oval-shaped progenitor cells in adult human livers that can be manipulated in culture to form either mature hepatocytes—cells that produce bile and break down toxins—or the epithelial cells that line bile ducts.

Generating "universal donor" cell lines would be another approach. To make such cells, scientists would remove, or use other molecules to mask, proteins on the surfaces of cells that normally identify the donor cells as "nonself." This strategy is now being used by Diacrin in Charlestown, Massachusetts, to make some types of pig cells acceptable for transplantation in humans. Diacrin also plans to use the "masking" technology to allow cell transplants between unmatched human donors. It has received regulatory approval in the U.S. to begin human trials of masked human liver cells for some cases of liver failure.

In principle, such universal donor cells would not be expected to be rejected by the recipient; they could be generated for various types of cells from many different tissues and kept growing in culture until needed. But it is not yet clear how universal donor cells will perform in large-scale clinical trials.

Parts Factories

Finding the best ways to produce cells and tissues has been far from straightforward. Scientists have identified only a handful of the biochemical signals that dictate the differentiation of embryonic stem cells and progenitor cells into specialized cell types, and we cannot yet isolate cultures of stem cells and progenitor cells from bone marrow without having connective tissue cells such as fibroblasts mixed in. (Fibroblasts are undesirable because they divide quickly and can overgrow cultures of stem cells.)

In addition, scientists need to develop more advanced procedures for growing cells in large quantities in so-called bioreactors, growth chambers equipped with stirrers and sensors that regulate the appropriate amounts of nutrients, gases such as oxygen and carbon dioxide, and waste products. Existing methods often yield too few cells or sheets of tissue that are thinner than desired.

New solutions are beginning to appear, however. For several years, researchers struggled to grow segments of cartilage that were thick enough for medical uses such as replacing worn-out cartilage in the knee. But once the cartilage grew beyond a certain thickness, the chondrocytes in the center were too far away from the growth medium to take up nutrients and gases, respond to growth-regulating chemical and physical signals, or expel wastes. Gordana Vunjak-Novakovic and Lisa Freed of the Massachusetts Institute of Technology solved the problem by culturing chondrocytes on a three-dimensional polymer scaffold in a bioreactor. The relatively loose weave of the scaffold and the stirring action of the bioreactor ensured that all the cells became attached uniformly throughout the scaffold material and were bathed in culture medium. Maximizing the mechanical properties of tissues as they grow in bioreactors will be crucial because many tissues remodel, or change their overall organization, in response to being stretched, pulled or compressed. Tissue-engineered cartilage, for example, becomes larger and contains more collagen and other proteins that form a suitable extracellular matrix if it is cultured in rotating vessels that expose the developing tissue to variations in fluid forces. (An extracellular matrix is a weblike network that serves as a support for cells to grow on and organize into tissues.) Cartilage cultured in this way contains extracellular matrix proteins that make it stiffer, more durable and more responsive physiologically to external forces.

Likewise, John A. Frangos of the University of California at San Diego has shown that osteoblasts cultured on a base of collagen beads being stirred in a bioreactor make more bone minerals than they do when they are grown in a flat, stationary dish. And Laura E. Niklason, who is now at Duke University, has demonstrated that tissue-engineered small arteries made of endothelial cells (blood vessel lining) and smooth muscle cells shaped into tubes develop mechanical properties more akin to natural blood vessels if they have growth medium pulsed through them to imitate the blood pressure generated by a beating heart. Several other teams—including ours—are developing ways to grow skeletal and cardiac muscle, tissues that become stronger as they respond to physical stress.

Desirable Properties

Learning how to regulate cell behavior represents another important challenge. Living systems are incredibly complex: The human liver, for example, contains six different types of cells that are organized into microscopic arrays called lobules. Each cell can perform hundreds of different biochemical reactions. What is more, the biochemical activity of each cell often depends on its interaction with other cells and with the network of extracellular matrix that wends through every tissue. David J. Mooney of the University of Michigan, for instance, has shown that hepatocytes produce varying levels of a given protein according to the stickiness of the material they are growing on. To develop organs such as an implantable, bioartificial liver—one major goal of tissue engineering—researchers must better understand how to grow hepatocytes and other cells of the liver under conditions that maximize their abilities to perform their normal physiological roles.

Understanding "remodeling" will be essential for crafting bioartificial organs and tissues that become a permanent part of the recipient. In the most successful laboratory tests of tissue-engineered products, the transplant has stimulated the growth of the recipient's own cells and tissues, which have eventually replaced the artificial polymers and transplanted cells of the graft. In collaboration with Toshiharu Shinoka and John E. Mayer of Children's Hospital in Boston, for instance, we have shown that a heart valve leaflet made of artificial polymers and lamb epithelial cells and myofibroblasts (a type of cell that helps to close wounds) became stronger, more elastic and thinner once transplanted into sheep. Moreover, the leaflet no longer consisted of artificial polymers after 11 weeks: it had been remodeled to contain only sheep extracellular matrix. Still, the precise biochemical signals and growth factors that dictate such remodeling processes remain essentially unknown.

Creating new materials that are biodegradable and do not induce the formation of scar tissue is an emerging area of tissue engineering that offers many challenges. Most of the materials now used as scaffolds for tissue engineering fall into one of two categories: synthetic materials such as biodegradable suture material or natural materials such as collagen or alginate (a gel-like substance derived from algae). The advantage of synthetic materials is that their strength, speed of degradation, microstructure and permeability can be controlled during production; natural materials, however, are usually easier for cells to stick to.

Researchers are now trying to combine the best of both worlds to design new generations of materials with particularly desirable properties. Some, for instance, are constructing biodegradable polymers that contain regions with biological activities that mimic the natural extracellular matrix of a particular tissue. One such polymer contains RGD, part of the extracellular matrix protein fibronectin. RGD is named for the single-letter abbreviation for the amino acids it is made of: arginine (R), glycine (G) and asparagine (D). Many types of cells normally stick to fibronectin by binding to RGD, so RGD-containing polymers might provide a more natural environment for growing cells.

Other scientists are attempting to make polymers that conduct electricity, which might be useful in growing tissue-engineered nerves, or polymers that gel rapidly. Such fast-setting polymers could be useful in injectable bioartificial products, including those that might be used to fill in a broken bone.

Inducing the growth of blood vessels, a process known as angiogenesis, will be key to sustaining many tissue-engineered organs—particularly pancreases, livers and kidneys, which require a large blood supply. Researchers have already successfully stimulated angiogenesis in bioartificial tissues growing in the laboratory by coating the polymer scaffolding supporting the tissues with growth factors that trigger blood vessel formation. Future studies will need to examine the best ways for releasing the growth factors and controlling their activity so that blood vessels form only when and where they are needed.

To the Patient

Developing new methods of tissue preservation is important to ensure that tissue-engineered products survive the trip from the factory to the operating room in good working order and do not die during transplantation. Technologies adapted from the field of donor-organ transplantation might be useful in this situation. For example, surgeons now know that much of the injury to a transplanted organ occurs during reperfusion, when the organ is connected to a blood supply in the recipient. Reperfusion induces the formation of oxygen free radicals, which literally poke holes in cell membranes and kill cells. To avoid reperfusion injury, surgeons currently add to the preservation solution chemicals that sop up such free radicals. Finding better molecules to protect tissue-engineered products from reperfusion injury and ischemic injury, which results when blood flow is insufficient, will be necessary. Cryopreservation techniques also need to be perfected so that bioartificial organs and tissues can be kept frozen until needed; methods currently used for cells will need to be developed further to work for larger tissues.

Determining the federal regulatory process for tissue-engineered products still presents a thorny issue. Bioartificial tissues and organs cut across nearly all the areas regulated by the U.S. Food and Drug Administration: they are essentially medical devices, but because they contain living cells they also produce biological substances that act like drugs. Accordingly, the FDA has treated the first tissue-engineered products to seek regulatory approval—two versions of bioartificial skin—as combination products. The agency is making tissue-engineering a priority area and is working to develop clear-cut policies to deal with bioartificial products.

We are confident that scientists and government regulators will clear all the hurdles described in this article to bring a variety of tissue-engineered products to the market in the coming years. Much challenging work remains, but someday—perhaps many years from now—equipping patients with tissue-engineered organs and tissues may be as routine as coronary bypasses are today.

Suggested Reading

Langer, Robert, and Joseph P. Vacanti. "Tissue Engineering," *Science* 260 (May 1993): 920–926.

Langer, Robert, and Joseph P. Vacanti. "Artificial Organs," *Scientific American* 273, no. 3 (September 1995): 130–133.

Lysaght, M. J., N. A. P. Nguy, and K. Sullivan. "An Economic Survey of the Emerging Tissue Engineering Industry," *Tissue Engineering* 4, no. 3 (Fall 1998): 231–238.

Chapter 2

When Things Go Really Wrong

How Cancer Arises

Robert A. Weinberg

How cancer develops is no longer a mystery. During the past two decades, investigators have made astonishing progress in identifying the deepest bases of the process—those at the molecular level. These discoveries are robust: they will survive the scrutiny of future generations of researchers, and they will form the foundation for revolutionary approaches to treatment. No one can predict exactly when therapies targeted to the molecular alterations in cancer cells will find wide use, given that the translation of new understanding into clinical practice is complicated, slow and expensive. But the effort is now under way.

In truth, the term "cancer" refers to more than 100 forms of the disease. Almost every tissue in the body can spawn malignancies; some even yield several types. What is more, each cancer has unique features. Still, the basic processes that produce these diverse tumors appear to be quite similar. For that reason, I will refer in this article to "cancer" in generic terms, drawing on one or another type to illustrate the rules that seem to apply universally.

The 30 trillion cells of the normal, healthy body live in a complex, interdependent condominium, regulating one another's proliferation. Indeed, normal cells reproduce only when instructed to do so by other cells in their vicinity. Such unceasing collaboration ensures that each tissue maintains a size and architecture appropriate to the body's needs.

Cancer cells, in stark contrast, violate this scheme; they become deaf to the usual controls on proliferation and follow their own internal agenda for reproduction. They also possess an even more insidious property—the ability to migrate from the site where they began, invading nearby tissues and forming masses at distant sites in the body. Tumors composed of such malignant cells become more and more aggressive over time, and they become lethal when they disrupt the tissues and organs needed for the survival of the organism as a whole.

This much is not new. But over the past 20 years, scientists have uncovered a set of basic principles that govern the development of cancer. We now know that the cells in a tumor descend from a common ancestral cell that at one point—usually decades before a tumor becomes palpable—initiated a program of inappropriate reproduction. Further, the malignant transformation of a cell comes about through the accumulation of mutations in specific classes of the genes within it. These genes provide the key to understanding the processes at the root of human cancer.

Genes are carried in the DNA molecules of the chromosomes in the cell nucleus. A gene specifies a sequence of amino acids that must be linked together to make a particular protein; the protein then carries out the work of the gene. When a gene is switched on, the cell responds by synthesizing the encoded protein. Mutations in a gene can perturb a cell by changing the amounts or the activities of the protein product.

Two gene classes, which together constitute only a small proportion of the full genetic set, play major roles in triggering cancer. In their normal configuration, they choreograph the life cycle of the cell—the intricate sequence of events by which a cell enlarges and divides. Proto-oncogenes encourage such growth, whereas tumor suppressor genes inhibit it. Collectively these two gene classes account for much of the uncontrolled cell proliferation seen in human cancers.

When mutated, proto-oncogenes can become carcinogenic oncogenes that drive excessive multiplication. The mutations may cause the proto-oncogene to yield too much of its encoded growth-stimulatory protein or an overly active form of it. Tumor suppressor genes, in contrast, contribute to cancer when they are inactivated by mutations. The resulting loss of functional suppressor proteins deprives the cell of crucial brakes that prevent inappropriate growth.

For a cancerous tumor to develop, mutations must occur in half a dozen or more of the founding cell's growth-controlling genes. Altered forms of yet other classes of genes may also participate in the creation of a malignancy, by specifically enabling a proliferating cell to become invasive or capable of spreading (metastasizing) throughout the body.

Signaling Systems Go Awry

Vital clues to how mutated proto-oncogenes and tumor suppressor genes contribute to cancer came from studying the roles played within the cell by the normal counterparts of these genes. After almost two decades of research, we now view the normal genetic functions with unprecedented clarity and detail.

Many proto-oncogenes code for proteins in molecular "bucket brigades" that relay growth-stimulating signals from outside the cell deep into its interior. The growth of a cell becomes deregulated when a mutation in one of its proto-oncogenes energizes a critical growth-stimulatory pathway, keeping it continuously active when it should be silent.

These pathways within a cell receive and process growth-stimulatory signals transmitted by other cells in a tissue. Such cell-to-cell signaling usually begins when one cell secretes growth factors. After release, these proteins move through the spaces between cells and bind to specific receptors—antennalike molecules—on the surface of other cells nearby. Receptors span the outer membrane of the target cells, so that one end protrudes into the extracellular space, and the other end projects into the cell's interior, its cytoplasm. When a growth-stimulatory factor attaches to a receptor, the receptor conveys a proliferative signal to proteins in the cytoplasm. These downstream proteins then emit stimulatory signals to a succession of other proteins, in a chain that ends in the heart of the cell, its nucleus. Within the nucleus, proteins known as transcription factors respond by activating a cohort of genes that help to usher the cell through its growth cycle.

Some oncogenes force cells to overproduce growth factors. Sarcomas and gliomas (cancers, respectively, of connective tissues and nonneuronal brain cells) release excessive amounts of platelet-derived growth factor. A number of other cancer types secrete too much transforming growth factor alpha. These factors act, as usual, on nearby cells, but, more important, they may also turn back and drive proliferation of the same cells that just produced them.

Researchers have also identified oncogenic versions of receptor genes. The aberrant receptors specified by these oncogenes release a flood of proliferative signals into the cell cytoplasm even when no growth factors are present to urge the cell to replicate. For instance, breast cancer cells often display Erb-B2 receptor molecules that behave in this way.

Still other oncogenes in human tumors perturb parts of the signal cascade found in the cytoplasm. The best-understood example comes from the *ras* family of oncogenes. The proteins encoded by normal *ras* genes transmit stimulatory signals from growth factor receptors to other proteins farther down the line. The proteins encoded by mutant *ras* genes, however, fire continuously, even when growth factor receptors are not prompting them. Hyperactive *ras* proteins are found in about a quarter of all human tumors, including carcinomas of the colon, pancreas and lung. (Carcinomas are by far the most common forms of cancer; they originate in epithelial cells, which line the body cavities and form the outer layer of the skin.) Yet other oncogenes, such as those in the *myc* family, alter the activity of transcription factors in the nucleus. Cells normally manufacture *myc* transcription factors only after they have been stimulated by growth factors impinging on the cell surface. Once made, *myc* proteins activate genes that force cell growth forward. But in many types of cancer, especially malignancies of the blood-forming tissues, *myc* levels are kept constantly high even in the absence of growth factors.

Discovery of trunk lines that carry proliferative messages from the cell surface to its nucleus has been more than intellectually satisfying. Because these pathways energize the multiplication of malignant cells, they constitute attractive targets for scientists intent on developing new types of anticancer therapeutics. In an exciting turn of events, as many as half a dozen pharmaceutical companies are working on drugs designed to shut down aberrantly firing growth factor receptors. At least three other companies are attempting to develop compounds that block the synthesis of aberrant *ras* proteins. Both groups of agents halt excessive signaling in cultured cancer cells, but their utility in blocking the growth of tumors in animals and humans remains to be demonstrated.

Tumor Suppressors Stop Working

To become malignant, cells must do more than overstimulate their growth-promoting machinery. They must also devise ways to evade or ignore braking signals issued by their normal neighbors in the tissue. Inhibitory messages received by a normal cell flow to the nucleus much as stimulatory signals do—via molecular bucket brigades. In cancer cells, these inhibitory brigades may be disrupted, thereby enabling the cell to ignore normally potent inhibitory signals at the surface. Critical components of these brigades, which are specified by tumor suppressor genes, are absent or inactive in many types of cancer cells.

A secreted substance called transforming growth factor beta (TGF-B) can stop the growth of various kinds of normal cells. Some colon cancer cells become oblivious to TGF-B by inactivating a gene that encodes a surface receptor for this substance. Some pancreatic cancers inactivate the *DPC4* gene, whose protein product may operate downstream of the growth factor receptor. And a variety of cancers discard the *p15* gene, which codes for a protein that, in response to signals from TGF-B, normally shuts down the machinery that guides the cell through its growth cycle.

Tumor suppressor proteins can also restrain cell proliferation in other ways. Some, for example, block the flow of signals through growth-stimulatory circuits. One such suppressor is the product of the *NF-1* gene. This cytoplasmic molecule ambushes the *ras* protein before it can emit its growth-promoting directives. Cells lacking *NF-1*, then, are missing an important counterbalance to *ras* and to unchecked proliferation.

Various studies have shown that the introduction of a tumor suppressor gene into cancer cells that lack it can restore a degree of normalcy to the cells. This response suggests a tantalizing way of combating cancer—by providing cancer cells with intact versions of tumor suppressor genes they lost during tumor development. Although the concept is attractive, this strategy is held back by the technical difficulties still encumbering gene therapy for many diseases. Current procedures fail to deliver genes to a large proportion of the cells in a tumor. Until this logistical obstacle is surmounted, the use of gene therapy to cure cancer will remain a highly appealing but unfulfilled idea.

The Clock Is Struck

Over the past five years, impressive evidence has uncovered the destination of stimulatory and inhibitory pathways in the cell. They converge on a molecular apparatus in the cell nucleus that is often referred to as the cell cycle clock. The clock is the executive decision maker of the cell, and it apparently runs amok in virtually all types of human cancer. In the normal cell, the clock integrates the mixture of growth-regulating signals received by the cell and decides whether the cell should pass through its life cycle. If the answer is positive, the clock leads the process.

The cell cycle is composed of four stages. In the G1 (gap 1) phase, the cell increases in size and prepares to copy its DNA. This copying occurs in the next stage, termed S (for synthesis), and enables the cell to duplicate precisely its complement of chromosomes. After the chromosomes are replicated, a second gap period, termed G2, follows, during which the cell prepares itself for M (mitosis)—the time when the enlarged parent cell finally divides in half to produce its two daughters, each of which is endowed with a complete set of chromosomes. The new daughter cells immediately enter G1 and may go through the full cycle again. Alternatively, they may stop cycling temporarily or permanently.

The cell cycle clock programs this elaborate succession of events by means of a variety of molecules. Its two essential components, cyclins and cyclin-dependent kinases (CDKs), associate with one another and initiate entrance into the various stages of the cell cycle. In G1, for instance, D-type cyclins bind to CDKs 4 or 6, and the resulting complexes act on a powerful growth-inhibitory molecule—the protein known as pRB. This action releases the braking effect of pRB and enables the cell to progress into late G1 and thence into S (DNA synthesis) phase.

Various inhibitory proteins can restrain forward movement through the cycle. Among them are p15 (mentioned earlier) and p16, both of which block the activity of the CDK partners of cyclin D, thus preventing the advance of the cell from G1 into S. Another inhibitor of CDKs, termed p21, can act throughout the cell cycle. P21 is under control of a tumor suppressor protein, p53, that monitors the health of the cell, the integrity of its chromosomal DNA and the successful completion of the different steps in the cycle.

Breast cancer cells often produce excesses of cyclin D and cyclin E. In many cases of melanoma, skin cells have lost the gene encoding the braking protein p16. Half of all types of human tumors lack a functional p53 protein. And in cervical cancers triggered by infection of cells with a human papilloma virus, both the pRB and p53 proteins are frequently disabled, eliminating two of the clock's most vital restraints. The end result in all these cases is that the clock begins to spin out of control, ignoring any external warnings to stop. If investigators can devise ways to impose clamps on the cyclins and CDKs active in the cell cycle, they may be able to halt cancer cells in their tracks.

I have so far discussed two ways that our tissues normally hold down cell proliferation and avoid cancer. They prevent excess multiplication by depriving a cell of growth-stimulatory factors or, conversely, by showering it with antiproliferative factors. Still, as we have seen, cells on their way to becoming cancerous often circumvent these controls: they stimulate themselves and turn a deaf ear to inhibitory signals. Prepared for such eventualities, the human body equips cells with certain backup systems that guard against runaway division. But additional mutations in the cell's genetic repertoire can overcome even these defenses and contribute to cancer.

Fail-Safe Systems Fail

One such backup system, present in each human cell, provokes the cell to commit suicide (undergo "apoptosis") if some of its essential components are damaged or if its control systems are deregulated. For example, injury to chromosomal DNA can trigger apoptosis. Further, recent work from a number of laboratories indicates that creation of an oncogene or the disabling of a tumor suppressor gene within a cell can also induce this response. Destruction of a damaged cell is bad for the cell itself but makes sense for the body as a whole: the potential dangers posed to the organism by carcinogenic mutations are far greater than the small price paid in the loss of a single cell. The tumors that emerge in our tissues, then, would seem to arise from the rare, genetically disturbed cell that somehow succeeds in evading the apoptotic program hardwired into its control circuitry.

Developing cancer cells devise several means of evading apoptosis. The p53 protein, among its many functions, helps to trigger cell suicide; its inactivation by many tumor cells reduces the likelihood that genetically troubled cells will be eliminated. Cancer cells may also make excessive amounts of the protein Bcl-2, which wards off apoptosis efficiently.

Recently scientists have realized that this ability to escape apoptosis may endanger patients not only by contributing to the expansion of a tumor but also by making the resulting tumors resistant to therapy. For years, it was assumed that radiation therapy and many chemotherapeutic drugs killed malignant cells directly, by wreaking widespread havoc in their DNA. We now know that the treatments often harm DNA to a relatively minor extent. Nevertheless, the affected cells perceive that the inflicted damage cannot be repaired easily, and they actively kill themselves. This discovery implies that cancer cells able to evade apoptosis will be far less responsive to treatment. By the same token, it suggests that therapies able to restore a cell's capacity for suicide could combat cancer by improving the effectiveness of existing radiation and chemotherapeutic treatment strategies.

A second defense against runaway proliferation, quite distinct from the apoptotic program, is built into our cells as well. This mechanism counts and limits the total number of times cells can reproduce themselves.

Cells Become Immortal

Much of what is known about this safeguard has been learned from studies of cells cultured in a petri dish. When cells are taken from a mouse or human embryo and grown in culture, the population doubles every day or so. But after a predictable number of doublings—50 to 60 in human cells—growth stops, at which point the cells are said to be senescent. That, at least, is what happens when cells have intact RB and p53 genes. Cells that sustain inactivating mutations in either of these genes continue to divide after their normal counterparts enter senescence. Eventually, though, the survivors reach a second stage, termed crisis, in which they die in large numbers. An occasional cell in this dying population, however, will escape crisis and become immortal: it and its descendants will multiply indefinitely.

These events imply the existence of a mechanism that counts the number of doublings through which a cell population has passed. During the past several years, scientists have discovered the molecular device that does this counting. DNA segments at the ends of chromosomes, known as telomeres, tally the number of replicative generations through which cell populations pass and, at appropriate times, initiate senescence and crisis. In so doing, they circumscribe the ability of cell populations to expand indefinitely [see "Telomeres, Telomerase and Cancer," by Carol W. Greider and Elizabeth H. Blackburn; *Scientific American*, February 1996].

Like the plastic tips on shoelaces, the telomere caps protect chromosomal ends from damage. In most human cells, telomeres shorten a bit every time chromosomes are replicated during the S phase of the cell cycle. Once the telomeres shrink below some threshold length, they sound an alarm that instructs cells to enter senescence. If cells bypass senescence, further shrinkage of the telomere will eventually trigger crisis: extreme shortening of the telomeres will cause the chromosomes in a cell to fuse with one another or to break apart, creating genetic chaos that is fatal to the cell.

If the telomere-based counting system operated properly in cancerous cells, their excessive proliferation would be aborted long before tumors became very large. Dangerous expansion would be stemmed by the senescence program or, if the cell evaded that blockade, by disruption of the chromosomal array at crisis. But this last defense is breached during the development of most cancer cells, overcome by activation of a gene that codes for the enzyme telomerase.

This enzyme, virtually absent from most healthy cell types but present in almost all tumor cells, systematically replaces telomeric segments that are usually trimmed away during each cell cycle. In so doing, it maintains the integrity of the telomeres and thereby enables cells to replicate endlessly. The resulting cell immortality can be troublesome in a couple of ways. Obviously, it allows tumors to grow large. It also gives precancerous or already cancerous cells time to accumulate additional mutations that will increase their ability to replicate, invade and ultimately metastasize.

From the point of view of a cancer cell, production of a single enzyme is a clever way to topple the mortality barrier. Yet dependence on one enzyme may represent an Achilles' heel as well. If telomerase could be blocked in cancer cells, their telomeres would once again shrink whenever they divided, pushing these cells into crisis and death. For that reason, a number of pharmaceutical firms are attempting to develop drugs that target telomerase.

Why Some Cancers Appear Early

It normally takes decades for an incipient tumor to collect all the mutations required for its malignant growth. In some individuals, however, the time for tumor development is clearly compressed; they contract certain types of cancer decades before the typical age of onset of these cancers. How can tumor formation be accelerated?

In many cases, this early onset is explained by the inheritance from one or the other parent of a mutant cancer-causing gene. As a fertilized egg begins to divide and replicate, the set of genes provided by the sperm and egg is copied and distributed to all the body's cells. Now a typically rare event—a mutation in a critical growth-controlling gene—becomes ubiquitous, because the mutation is implanted in all the body's cells, not merely in some randomly stricken cell. In other words, the process of tumor formation leapfrogs over one of its early, slowly occurring steps, accelerating the process as a whole. As a consequence, tumor development, which usually requires three or four decades to reach completion, may culminate in one or two. Because such mutant genes can pass from generation to generation, many members of a family may be at risk for the early development of cancer.

An inherited form of colon cancer provides a dramatic example. Most cases of colon cancer occur sporadically, the results of random genetic events occurring during a person's lifetime. In certain families, however, many individuals are afflicted with early-onset colonic tumors, preordained by an inherited gene. In the sporadic cases, a rare mutation silences a tumor suppressor gene called APC in an intestinal epithelial cell. The resulting proliferation of the mutant cell yields a benign polyp that may eventually progress to a malignant carcinoma. But defective forms of APC may pass from parents to children in certain families. Members of these families develop hundreds, even thousands of colonic polyps during the first decades of life, some of which are likely to be transformed into carcinomas.

The list of familial cancer syndromes that are now traceable directly to inheritance of mutant tumor suppressor genes is growing. For instance, inherited defective versions of the gene for *pRB* often lead to development of an eye cancer—retinoblastoma—in children; later in life the mutations account for a greatly increased risk of osteosarcomas (bone cancers). Mutant inherited versions of the *p53* tumor suppressor gene yield tumors at multiple sites, a condition known as the Li-Fraumeni syndrome. And the recently isolated *BRCA1* and *BRCA2* genes seem to account for the bulk of familial breast cancers, encompassing as many as 20 percent of all premenopausal breast cancers in this country and a substantial proportion of familial ovarian cancers as well.

Early onset of tumors is sometimes explained by inheritance of mutations in another class of genes as well. As I implied earlier, most people avoid cancer until late in life or indefinitely because they enter the world with pristine genes. During the course of a lifetime, however, our genes are attacked by carcinogens imported into our bodies from the environment and also by chemicals produced in our own cells. And genetic errors may be introduced when the enzymes that replicate DNA during cell cycling make copying mistakes. For the most part, such errors are rapidly corrected by a repair system that operates in every cell. Should the repair system slip up and fail to erase an error, the damage will become a permanent mutation in one of the cell's genes and in that same gene in all descendant cells.

The system's high repair efficiency is one reason many decades can pass before all the mutations needed for a malignancy to develop will, by chance, come together within a single cell. Certain inherited defects, though, can accelerate tumor development through a particularly insidious means: they impair the operation of proteins that repair damaged DNA. As a result, mutations that would normally accumulate slowly will appear with alarming frequency throughout the DNA of cells. Among the affected genes are inevitably those controlling cell proliferation.

Such is the case in another inherited colon cancer, hereditary nonpolyposis colon cancer. Afflicted individuals make defective versions of a protein responsible for repairing the copying mistakes made by the DNA replication apparatus. Because of this impairment, colonic cells cannot fix DNA damage efficiently; they therefore collect mutations rapidly, accelerating cancer development by two decades or more. People affected by another familial cancer syndrome, xeroderma pigmentosum, have inherited a defective copy of a gene that directs the repair of DNA damaged by ultraviolet rays. These patients are prone to several types of sunlight-induced skin cancer.

Similarly, cells of people born with a defective *ATM* gene have difficulty recognizing the presence of certain lesions in the DNA and mobilizing the appropriate repair response. These people are susceptible to neurological degeneration, blood vessel malformation and a variety of tumors. Some researchers have proposed that as many as 10 percent of inherited breast cancers may arise in patients with a defective copy of this gene.

Over the next decade, the list of cancer susceptibility genes will grow dramatically, one of the fruits of the Human Genome Project (which seeks to identify every gene in the human cell). Together with the increasingly powerful tools of DNA analysis, knowledge of these genes will enable us to predict which members of cancer-prone families are at high risk and which have, through good fortune, inherited intact copies of these genes.

Beyond Proliferation

Although we have learned an enormous amount about the genetic basis of runaway cell proliferation, we still know rather little about the mutant genes that contribute to later stages of tumor development, specifically those that allow tumor cells to attract blood vessels for nourishment, to invade nearby tissues and to metastasize. But research in these areas is moving rapidly.

We are within striking distance of writing the detailed life histories of many human tumors from start to life-threatening finish. These biographies will be written in the language of genes and molecules. Within a decade, we will know with extraordinary precision the succession of events that constitute the complex evolution of normal cells into highly malignant, invasive derivatives.

By then, we may come to understand why certain localized masses never progress beyond their benign, noninvasive form to confront us with aggressive malignancy. Such benign growths can be found in almost every organ of the body. Perhaps we will also discern why certain mutant genes contribute to the formation of some types of cancer but not others. For example, mutant versions of the *RB* tumor suppressor gene appear often in retinoblastoma, bladder carcinoma and small cell lung carcinoma but are seen only occasionally in breast and colon carcinomas. Very likely, many of the solutions to these mysteries will flow from research in developmental biology (embryology). After all, the genes that govern embryonic development are, much later, the sources of our malignancies.

By any measure, the amount of information gathered over the past two decades about the origins of cancer is without parallel in the history of biomedical research. Some of this knowledge has already been put to good use, to build molecular tools for detecting and determining the aggressiveness of certain types of cancer. Still, despite so much insight into cause, new curative therapies have so far remained elusive. One reason is that tumor cells differ only minimally from healthy ones; a minute fraction of the tens of thousands of genes in a cell suffers damage during malignant transformation. Thus, normal friend and malignant foe are woven of very similar cloth, and any fire directed against the enemy may do as much damage to normal tissue as to the intended target.

Yet the course of the battle is changing. The differences between normal and cancer cells may be subtle, but they are real. And the unique characteristics of tumors provide excellent targets for intervention by newly developed drugs. The development of targeted anticancer therapeutics is still in its infancy. This enterprise will soon move from hit-or-miss, serendipitous discovery to rational design and accurate targeting. I suspect that the first decade of this century will reward us with cancer therapies that earlier generations could not have dreamed possible. Then this nation's long investment in basic cancer research will begin to pay off handsomely.

Outbreak Not Contained

Marguerite Holloway

The appearance of West Nile virus in New York City in summer 1999 caught the U.S. by surprise. That this virus—which is known in Africa, Asia and, increasingly, parts of Europe—could find its way to American shores and perform its deadly work for many months before being identified has shaken up the medical community. It has revealed several major gaps in the public health infrastructure that may become ever more important in this era of globalization and emerging diseases.

Because it is mosquito-borne, West Nile has reinforced the need for mosquito surveillance—something that is only sporadically practiced around the country and something that could perhaps help doctors identify other agents causing the many mysterious cases of encephalitis that occur every year. And because it killed birds before it killed seven people, the virus made dramatically clear that the cultural divide between the animal-health and the public-health communities is a dangerous one. "It was a tremendous wake-up call for the United States in general," says William K. Reisen of the Center for Vector-Borne Disease Research at the University of California at Davis.

No one is certain when, or how, West Nile arrived in New York. The virus—one of 10 in a family called flaviviruses, which includes St. Louis encephalitis—could have come via a bird, a mosquito that had survived an intercontinental flight or an infected traveler. It is clear, however, that West Nile started felling crows in New York's Queens County in June and had moved into the Bronx by July, where it continued to kill crows and then, in September, birds at the Bronx Zoo.

By the middle of August, people were succumbing as well. In two weeks Deborah S. Asnis, chief of infectious disease at the Flushing Hospital Medical Center in Queens, saw eight patients suffering similar neurological complaints. After the third case, and despite some differences in their symptoms, Asnis alerted the New York City Department of Health. The health department, in turn, contacted the state and the Centers for Disease Control and Prevention (CDC), and the hunt for the pathogen was on. It was first identified as St. Louis encephalitis, which has a similar clinical profile and cross-reacts with West Nile in the lab.

Understandable as it is to many health experts, the initial misidentification remains worrisome. As Reisen points out, diagnostic labs can only look for what they know. If they don't have West Nile reagents on hand, they won't find the virus, just its relatives. "In California we have had only one flavivirus that we were looking for, so if West Nile had come in five years ago, we would have missed it until we had an isolate of the virus as well," Reisen comments.

This is true even though California, unlike New York State, has an extensive, $70 million-a-year mosquito surveillance and control system. The insects are trapped every year so that their populations can be assessed and tested for viruses. Surveillance has allowed California to document the appearance of three new species of mosquito in the past 15 years. In addition, 200 flocks of 10 sentinel chickens are stationed throughout the state. Every few weeks during the summer they are tested for viral activity.

In 1990 sentinel chickens in Florida detected St. Louis encephalitis before it infected people. "Six weeks before the human cases, we knew we had a big problem," recalls Jonathan F. Day of the Florida Medical Entomology Laboratory. After warning people to take precautions and spraying with insecticides, the state documented 226 cases and 11 deaths: "It is very difficult to say how big the problem would have been if we hadn't known," Day says. "But without our actions I think it would have been in the thousands." (Day says surveillance in his county costs about $35,000 annually.)

New York City, home to perhaps about 40 species of mosquito, has no such surveillance in place, even though some of its neighbors—Suffolk County, Nassau County and every county in New Jersey—do. And it is not alone. Many cities do not monitor the whining pests unless they are looking for a specific disease. "We have clearly forgotten about mosquito-borne disease," says Thomas P. Monath, vice president of research and medical affairs at OraVax in Cambridge, Massachusetts, and formerly of the CDC. "We have let our infrastructure decay, and we have fewer experts than we had twenty or thirty years ago."

Tracking mosquitoes could potentially help not just with exotic arrivals but with the plethora of encephalitis cases reported every year. Indeed, the particular strain of West Nile that hit New York was ultimately identified by Ian Lipkin of the University of California at Irvine and his colleagues because they were collaborating with the New York State Department of Health on an encephalitis project. Two-thirds of the cases of encephalitis that occur every year have an "unknown etiology." A few states, including New York, California and Tennessee, have recently started working with the CDC to develop better tests to identify some of these mysterious origins. As a result, Lipkin—who says he has developed an assay that can quickly identify pathogens even if they are not being looked for—was given samples from the patients who had died in New York City.

Some health experts, including Mahfouz H. Zaki of the Suffolk County Department of Health Services, predict that better mosquito surveillance would lead to a better understanding of encephalitis in general. As Zaki has noted, most of the unknown-etiology cases occur in September—just when when insect-borne diseases tend to peak. Three hundred of the 700 such cases in New York State every year occur in New York City.

Even if surveillance can't catch what it doesn't know, it can tell public health researchers that a new mosquito species has appeared—say, one that can transmit dengue or yellow fever—or it can indicate that something is wrong with the birds and should be investigated. The sentinels in the case of West Nile were, in fact, the city's crows and, later, birds at the Bronx Zoo. Through careful analysis of the crows and other species, Tracey McNamara, a veterinary pathologist at the Wildlife Conservation Society (which runs the Bronx Zoo), quickly determined that the pathogen was not St. Louis encephalitis—despite the CDC claims—because that disease does not kill birds. And she knew that it was not eastern equine encephalitis, because emus weren't dying. "We owe a debt of gratitude to the emu flock," McNamara says.

But despite her recognition that something new, unusual and deadly was afoot, McNamara could do little herself—except hound people in the human-health community to take a look at the wildlife. "The thing that was so frustrating was that we lack the infrastructure to respond," she says. "There was no vet lab in the country that could do the testing." Because none of the veterinary or wildlife labs had the ability to deal with such pathogens, McNamara was forced to send her bird samples to the CDC and to a U.S. Army lab. The Wildlife Conservation Society recently gave $15,000 to Robert G. McLean, director of the U.S. Geological Survey's National Wildlife Health Center, so he could study the pathogenesis of West Nile virus in crows and the effectiveness of an avian vaccine. "The federal budget moves at glacial speed," McNamara complains. "That is going to need to be addressed."

The continued bird work by McLean and others has kept the East Coast on alert for the potential of another West Nile outbreak this summer. In the fall of 1999 McLean and his colleagues found West Nile in a crow in Baltimore and in a migratory bird, the eastern phoebe. "They go to the southern U.S.," he notes. "That just convinces us that a lot of migratory birds were infected and flew south with the virus." Despite the fact that "wildlife is a good warning system for what could eventually cause problems in humans," McLean is not optimistic about a true and equal collaboration between his and McNamara's world and the CDC's: "We are on the outside looking in. We are not partners yet, and I am not sure we will ever get to be partners." The cost could be high. As McNamara points out, "Don't you want a diagnosis in birds before it gets to humans?"

Invasion of the Body Snatchers

Carol Ezzell

If being pregnant feels like an alien has invaded your body, there's a good reason for it: essentially, that's just what happens. Researchers have long pondered how a mother's immune system learns to tolerate the presence of an embryo or fetus—at least half of which (the paternal half, that is) is foreign to the mother.

The mother's immune system has to ignore the nine-month embryonic squatter as it grows. (Otherwise, a quick eviction in the form of a miscarriage usually ensues.) But such maternal tolerance can backfire, scientists now say.

Indeed, new evidence suggests that fetal cells left over from pregnancy can linger in a woman's body for decades. When these cells are remnants of the fetus's own developing immune system, they can cause diseases in the mother that were previously thought to be autoimmune, in which the individual's immune system attacks his or her other tissues.

In the April 23, 1998 issue of the *New England Journal of Medicine*, Sergio A. Jimenez and his colleagues at Thomas Jefferson University in Philadelphia report that they found fetal immune cells in skin lesions taken from women with the disorder systemic sclerosis. Taken together with previous studies, the report is shaking up the standard view among scientists of what causes autoimmune diseases.

Systemic sclerosis affects roughly 150,000 people in the U.S.—mostly women between the ages of 30 and 60. People with a subset of the disease called scleroderma (meaning hard skin) produce excess connective tissue called collagen and develop thickened, taut skin as result. For some, the thickenings are only of cosmetic concern, but in others they restrict movement. In patients with systemic sclerosis, the thickening extends to joints, arteries and internal organs, such as the gastrointestinal tract, lungs, kidneys and heart. Many people with the more aggressive forms of systemic sclerosis die of cardiopulmonary or kidney failure within five years of diagnosis; only about 50 percent live for 10 years after their diagnosis.

Researchers have been studying the link between fetal cells and autoimmune disease since 1996, when Diana W. Bianchi of Tufts University and her colleagues reported that not only can women carry fetal cells in their blood for as long as 27 years after pregnancy, but many actually do. Bianchi and her coworkers identified fetal cells in women who had given birth to sons. Because women do not have Y chromosomes, it was easy for the researchers to detect the male fetal cells in the mother's blood simply by looking for DNA sequences from the Y chromosome.

In February of 1998 J. Lee Nelson of the Fred Hutchinson Cancer Research Center in Seattle and her colleagues used the same technique to tie fetal cells to scleroderma. In a paper published in the journal the *Lancet*, Nelson's group reported that women with scleroderma had more than 10 times as many fetal cells in their blood as women without the disease.

Nelson and her coworkers concluded that scleroderma is analogous to graft-versus-host disease (GVHD), a possible complication of bone marrow transplantation. In GVHD, immune cells called T-cells from the donor bone marrow attack the tissues of the recipient, causing some of the same symptoms as those seen in systemic sclerosis.

Not everyone accepts that idea. In an editorial accompanying Nelson's *Lancet* paper, Ken Walsh of Churchill Hospital in Oxford, England, called the similarities between scleroderma and GVHD "peripheral." "There is no proof that any [bone marrow] transplant recipient ever developed scleroderma," he wrote.

But the report by Jimenez's group strengthens the apparent link between systemic sclerosis and GVHD. Like Nelson's group, the researchers found Y chromosome DNA sequences in the blood of 32 of 69 women with systemic sclerosis and in only 1 of 25 healthy women. Jimenez and his colleagues also found Y chromosome sequences in the skin lesions of 11 of 19 affected women.

Jimenez says that he and his colleagues have early indications that the fetal cells containing the Y chromosome sequences are T-helper cells, which secrete substances called cytokines. Cytokines are chemical messengers that activate the immune system. He speculates that these cells arise from immature fetal stem cells, which cross over into the mother's circulation and then mature. "It's still pretty much a hypothesis that needs to be explored," Jimenez adds. "We need to see how mature the cells we've found are."

If Jimenez and his coworkers are right, it could revolutionize therapy for systemic sclerosis and scleroderma, which now consists of nothing more than blood-pressure-lowering drugs to slow kidney damage. Jimenez says that if future studies confirm that fetal T-helper cells are to blame, researchers might one day use cytokine-blocking drugs to treat the diseases. And if they can identify distinguishing molecules on the surfaces of the fetal cells, those molecules could serve as the basis for vaccines that would wake up the mother's immune system and sic it on the foreign cells.

The new study doesn't explain why all women with fetal cells in their circulation—however many that might be—don't get systemic sclerosis. Jimenez acknowledges that environmental factors, such as exposure to toxins, must also play a role. "You must have a second event to activate the fetal T-cells," he says. "The mere presence of the cells can't account for the disease."

Nor does the study suggest what might cause systemic sclerosis among the relatively few men who develop the disease, and exactly how the fetal T-helper cells cause the overproduction of collagen that leads to the skin thickening and other symptoms of systemic sclerosis.

Nevertheless, Jimenez is optimistic that researchers have turned a corner when it comes to understanding systemic sclerosis and scleroderma. "For many years, we've been going in circles studying scleroderma," Jimenez says. "Now it looks like we might just be getting somewhere."

Deadly Enigma

Tim Beardsley

It is, in the words of one group of researchers, "a true quandary." How can an abnormal form of a protein present in all mammals cause some 15 different lethal brain diseases that affect animals as diverse as hamsters, sheep, cattle, cats and humans? Yet the dominant theory about the group of illnesses that includes scrapie in sheep, mad cow disease in cattle and Creutzfeldt-Jakob disease in humans holds just that. What is certain is that some mysterious agent that resists standard chemical disinfection as well as high temperatures can transmit these diseases between individuals and, less often, between species. What is unknown is how the agent spreads under natural conditions and how it destroys brain tissue. Because of the characteristic spongelike appearance of brain tissue from stricken animals, the diseases are called transmissible spongiform encephalopathies (TSEs).

Finding the answers is a matter of urgency. In Britain, mad cow disease, or bovine spongiform encephalopathy, has turned into a national calamity. A worldwide ban is on British beef and livestock imports. The government is slaughtering all cattle older than 30 months—some 30,000 a week—to allay fears that the disease, which causes animals to become nervous and develop an unsteady gait, will spread to people. So far British medical researchers have identified 14 unusual cases of Creutzfeldt-Jakob disease in young people that they suspect were a human manifestation of mad cow disease. New studies of the victims' brains appear to strengthen that conclusion. The biochemical properties of the suspected disease-causing protein in the brains of the victims are distinctly different from those usually found in Creutzfeldt-Jakob disease, supporting the notion that the disease came from a novel source.

Apprehensive that the U.S. cattle industry could be in line for a disaster like the one in Britain, in October 1996 the Food and Drug Administration was about to propose controls on the use of animal-derived protein and bone meal in cattle feed. Mad cow disease is believed to have spread in Britain because of the practice of incorporating material from the rendered carcasses of cattle and other animals into cattle feed. That cannibalistic practice is also standard in the U.S.

Although only one case of the disease has been confirmed in North America—in an animal imported from Britain to Canada—other TSEs, including scrapie in sheep and comparable diseases in mink and mule deer, are well known in the U.S. Nobody has any idea whether some native scrapielike agent could transform itself into mad cow disease or something unpleasantly like it. "As long as we continue to feed cows to cows we are at risk," says Richard F. Marsh of the University of Wisconsin, who has studied TSE in mink. The cattle-rendering industry, however, is resisting blanket bans and wants to see controls only on tissues for which there is firm evidence of infectivity.

Unfortunately, the science of TSEs generally is not in a firm state. Laboratory tests show that the diseases have variable and strange characteristics. They are most easily transmitted by injecting brain tissue from an infected animal into a recipient's brain, but sometimes eating brain or other offal will do the job. (Kuru, a human TSE formerly common in Papua New Guinea, was spread because the Fore people ritually consumed the brains of their dead.) There are distinct strains of some TSEs, including scrapie and Creutzfeldt-Jakob disease, but passage through a different species can permanently alter the diseases' pathological characteristics in the original host species.

The leading theory that ties these characteristics together comes from Stanley B. Prusiner of the University of California at San Francisco [see "The Prion Diseases," by Stanley B. Prusiner; *Scientific American*, January 1995]. The theory posits that a ubiquitous mammalian protein called prion protein can, rarely, refold itself into a toxic form that then spreads the conversion of more healthy protein in a runaway process. Some mutant forms of the protein are more likely to convert spontaneously than others, which accounts for rare sporadic cases. TSEs are thus both inherited and transmissible, and unlike those of any other known diseases, the pathogen lacks DNA or RNA.

Some of the strongest evidence for Prusiner's theory is his demonstration that mice genetically engineered to produce an abnormal prion protein develop a spongiform disease and can transmit illness to other mice via their brain tissue. Critics, such as Richard Rubenstein of the New York Institute for Basic Research, note that the mice in these experiments contain very little of the abnormal prion protein that is supposed to be the disease agent. So, Rubenstein argues, they may not be truly comparable to animals with TSEs. Perhaps, Rubenstein and others suggest, some toxin in the brains of the sick experimental mice caused the recipients of their tissue to become sick, too. Prusiner maintains, however, that no ordinary toxin is potent and slow enough to give his results.

Prusiner insists his most recent experiments, which employ elaborate tests designed to rule out possible sources of error, make his theory unassailable. And one of Prusiner's chief rivals, Byron W. Caughey of the Rocky Mountain Laboratories of the National Institutes of Health in Hamilton, Montana, has made the protein-only theory more plausible by experiments that he believes replicate the process by which TSEs propagate in the brain. Caughey and his associates have shown that under specific chemical conditions, they can convert some of the normal prion protein into the abnormal form in the test tube. Moreover, abnormal proteins from different strains of scrapie, which are chemically distinguishable, seem to produce their own strain-specific type of abnormal protein.

Caughey believes his experiments indicate that normal, healthy prion protein changes into the pathological variant when it forms aggregates of some 20 to 50 molecules. The process gets under way if it is seeded by a piece of the abnormal aggregate. Together with Peter T. Lansbury of the Massachusetts Institute of Technology, Caughey has proposed a geometric model illustrating that aggregates can form in different crystalline patterns corresponding to different TSEs.

Caughey says he is keeping an open mind on whether there might be some DNA or RNA along with the protein that might help explain the variety of TSEs. The ultimate proof of the protein-only theory would be to fabricate abnormal protein from simple chemicals and show that it caused transmissible disease in animals, but neither Caughey nor anyone else can do that. Caughey's experiments still need a seed from a sick animal, and the amount of abnormal protein the experiments produce is not enough to prove that the freshly created material can cause disease.

Prusiner, for his part, is not about to concede to Caughey. He believes aggregates are merely an artifact of Caughey's experimental procedures. "There are no ordered aggregates of polymers of prion protein in cells in the brain," he declares. Prusiner's studies lead him to think, instead, that an as yet unidentified "protein X" is responsible for converting the normal prion protein to the scrapie form. He and his coworkers have synthesized fragments of the healthy prion protein and shown that they can spontaneously form fibrils that resemble those seen in the TSE diseases.

Whether protein-only prions can explain TSEs or not, it will take more than a decade for British scientists to unravel how BSE spreads, predicts D. Carleton Gajdusek of the NIH, who first showed how kuru spreads. A test for TSEs in humans and in a few animals was announced in September 1996, but so far it seems to perform well only when clear symptoms of illness have already developed. Although the test may be useful to confirm suspected TSEs in humans, the most important step for governments to take, Gajdusek says, is to maintain intensive surveillance for patients with unusual neurological symptoms. His pictures and descriptions of children with kuru have been distributed to neurologists in Europe to help them recognize possible victims.

By the Numbers: Health Care Costs

Rodger Doyle

Rising medical costs are a worldwide problem, but nowhere are they higher than in the U.S. Although Americans with good health insurance coverage may get the best medical treatment in the world, the health of the average American, as measured by life expectancy and infant mortality, is below the average of other major industrial countries. Inefficiency, fraud and the expense of malpractice suits are often blamed for high U.S. costs, but the major reason is overinvestment in technology and personnel. America leads the world in expensive diagnostic and therapeutic procedures, such as organ transplants, coronary artery bypass surgery and magnetic resonance imaging. Orange County, California, for example, has more MRI machines than all of Canada.

Federal policy since World War II has emphasized medical technology and the widespread building of hospitals, even in rural areas. Other industrial countries, in contrast, followed the more cost-effective alternative of building up regional centers. The U.S. has long overinvested in the training of specialists at the expense of primary physicians, leading to a large surplus of specialists. Because specialists have economic incentives to perform unnecessary procedures, they may contribute to cost inflation.

Other industrial countries have managed to slow the growth in costs while achieving near-universal coverage. These include Britain, France and Italy, which have heavily centralized systems; Canada and Germany, which have decentralized systems but whose provinces play a key administrative role; and Japan, which combines strong national policy making with health care administration left largely in private hands. In each instance, central governments imposed strict fiscal controls even though they resulted in long waiting times for elective treatment and considerable delays in seeing specialists.

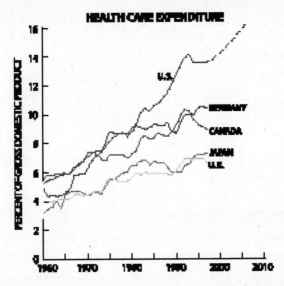

HEALTH CARE EXPENDITURE

Source: Organization for Economic Cooperation and Development, Health Data 1997. Dashed line shows projections for U.S. made by Sheila Smith, Mark Freeland, Stephen Heffler et al., "The Next Ten Years of Health Spending," in *Health Affairs* 17, no. 5 (September–October 1998), 128–140.

President Bill Clinton attempted to impose central fiscal controls as a part of his 1994 health care plan but was unable to put together a solid supporting coalition. Insurance firms, pharmaceutical companies, small business operators and academic medical centers were opposed to the plan. Labor unions and Medicare beneficiaries generally favored it but lobbied vigorously for changes that would improve their benefits. Republicans opposed the plan on the grounds that it called for new taxes.

According to political scientist Lawrence R. Jacobs of the University of Minnesota, universal access is a key to the success of other countries in imposing fiscal controls because it helps to lessen friction between groups. The American system encourages discord, for example, between health care insurers and high-risk people whom they exclude from coverage. Americans who receive adequate care through employers have little economic interest in seeing coverage extended to the more than 43 million Americans now uninsured.

In recent years U.S. health care expenditures as a percent of gross domestic product have leveled off, probably as a result of the expansion of managed care. The projected increase to 16.6 percent of GDP in 2007 shown on the chart assumes that managed care will grow more slowly, that increasing consumer income will boost the demand for medical services and that medical cost inflation will accelerate. But the period of greatest stress will come after 2010, when baby boomers begin to retire. Not only will federal budgets be strained, but also employers, already paying far more in medical costs than foreign competitors, will be put at a further disadvantage in world trade.

How can the federal government ever assert fiscal control over medical costs? Victor R. Fuchs of Stanford University, a longtime observer of the medical economy, believes that comprehensive reform of the U.S. medical system will come only after a major political crisis as might accompany war, depression or widespread civil unrest. Such a crisis might arise as medical costs reach ever higher and threaten Social Security, Medicare and other popular programs; there could be political upheaval of such magnitude that medical reform will seem to be the easy solution.

Mutations Galore

Tim Beardsley

All living things slowly accumulate mutations, changes in the string of chemical units in the famous DNA double helix that may in turn alter the form and function of a protein. A mutation that does affect a protein, if passed on to an offspring, might improve the progeny's chances in life—or, more likely, harm them. Deleterious mutations, which can cause genetic diseases, are unfortunately more likely than beneficial ones, for the same reason that randomly retuning a string on a piano is likely to make the instrument sound worse, not better.

Despite the hazard of harmful mutations, researchers until recently had only the vaguest notion of how often they occur in humans. Many mutations are thought to produce no obvious effect, yet they might still represent a subtle disadvantage to an organism carrying them. Adam Eyre-Walker of the University of Sussex and Peter D. Keightley of the University of Edinburgh recently examined the frequency of mutations in humans by studying how many have occurred in a sample of 46 genes during the six million years since humans and chimpanzees last shared an ancestor. The results, published in *Nature*, were surprising: a minimum of 1.6 harmful mutations occurs per person per generation, and the number is more likely close to three. That number is high enough to pose a challenge to theorists.

Eyre-Walker and Keightley's approach was subtle. They first assessed how many human mutations occurred in the sample of genes that could not have produced any alteration in a protein and so must have been invisible to natural selection. (A fair proportion of mutations, even those occurring in active genes, do not cause any change in the protein that they encode.) They judged which differences in gene sequences between humans and chimpanzees were caused by mutations in humans by comparing discrepant sequences with the equivalent gene sequence in a third primate group. If the third group's sequence matched up with that of the chimpanzees, the change was surmised to have occurred in the human line.

From this observed number of "invisible" human mutations, Eyre-Walker and Keightley could calculate the theoretical number of mutations that should have resulted in altered proteins. The answer was 231. But only 143 such protein-changing human mutations were actually seen in the sample. The missing 88, they concluded, did occur at some point but were harmful enough to be eliminated by natural selection. That number leads to the estimate of perhaps three harmful mutations per person per generation.

The proportion of mutations that is clearly harmful seems lower than most geneticists would have guessed. But the overall rate of human mutations is very high, and as a result the actual rate of bad mutations is disturbingly high, too.

According to standard population genetics theory, the figure of three harmful mutations per person per generation implies that three people would have to die prematurely in each generation (or fail to reproduce) for each person who reproduced, in order to eliminate the now absent deleterious mutations. Humans do not reproduce fast enough to support such a huge death toll. As James F. Crow of the University of Wisconsin asked rhetorically, in a commentary in *Nature* on Eyre-Walker and Keightley's analysis: "Why aren't we extinct?"

Crow's answer is that sex, which shuffles genes around, allows detrimental mutations to be eliminated in bunches. The new findings thus support the idea that sex evolved because individuals who (thanks to sex) inherit several bad mutations rid the gene pool of all of them at once, by failing to survive or reproduce.

Yet natural selection has weakened in human populations with the advent of modern medicine, Crow notes. So he theorizes that harmful mutations might now be starting to accumulate at an even greater rate, with possibly worrisome consequences for health. Keightley is skeptical: he thinks that many mildly deleterious mutations have already become widespread in human populations through random events in evolution and that various adaptations, notably intelligence, have more than compensated. "I doubt that we'll have to pay a penalty as Crow seems to think," he remarks. "We've managed perfectly well up until now."

Suggested Reading

Bishop, J. M. "Cancer: The Rise of the Genetic Pardigm," *Genes and Development* 9, no. 11 (June 1995): 1309–1315.

Cairns, J. *Cancer: Science and Society*. New York: W. H. Freeman, 1978.

Cooper, G. M. *Ongogenes*. 2nd ed. Boston: Jones and Bartlett Publishers, 1995.

Varmus, H., and R. A. Weinberg. *Genes and the Biology of Cancer*. New York: Scientific American Library (distributed by W. H. Freeman), 1993.

Vogelstein, B., and K. W. Kinzler. "The Multistep Nature of Cancer," *Trends in Genetics* 9, no. 4 (April 1993): 138–141.

Chapter 3

Using Biochemistry on the Defensive

Embryonic Stem Cells for Medicine

Roger A. Pedersen

Your friend has suffered a serious heart attack while hiking in a remote region of a national park. By the time he reaches a hospital, only one-third of his heart is still working, and he seems unlikely to return to his formerly active life. Always the adventurer, though, he volunteers for an experimental treatment. He provides a small sample of skin cells. Technicians remove the genetic material from the cells and inject it into donated human eggs from which the nucleus, which houses the gene-bearing chromosomes, has been removed. These altered eggs are grown for a week in a laboratory, where they develop into early-stage embryos. The embryos yield cells that can be cultured to produce what are called embryonic stem cells. Such cells are able to form heart muscle cells, as well as other cell types.

The medical team therefore establishes a culture of embryonic stem cells and grows them under conditions that induce them to begin developing into heart cells. Being a perfect genetic match for your friend, these cells can be transplanted into his heart without causing his immune system to reject them. They grow and replace cells lost during the heart attack, returning him to health and strength.

This scenario is for now hypothetical, but it is not fantastic. Researchers already know of various types of stem cells. These are not themselves specialized to carry out the unique functions of particular organs, such as the heart, the liver or the brain. But when stem cells divide, some of the progeny "differentiate"—that is, they undergo changes that commit them to mature into cells of specific types. Other progeny remain as stem cells. Thus, intestinal stem cells continually regenerate the lining of the gut, skin stem cells make skin, and hematopoietic stem cells give rise to the range of cells found in blood. Stem cells enable our bodies to repair everyday wear and tear.

Embryonic stem cells are even more extraordinary: they can give rise to essentially all cell types in the body. Human embryonic stem cells were first grown in culture in 1998. In February 1998 James A. Thomson of the University of Wisconsin found the first candidates when he noted that certain human cells plucked from a group growing in culture resembled embryonic stem cells that he had earlier derived from rhesus monkey embryos. A thousand miles away in Baltimore, John D. Gearhart of Johns Hopkins University was isolating similar cells by culturing fragments of human fetal ovaries and testes. And in California, researchers at Geron Corporation in Menlo Park and in my laboratory at the University of California at San Francisco were carrying out related studies.

But Thomson was well served by his previous experience with embryonic stem cells of rhesus monkeys and marmosets, which—like humans—are primates. In the following months he pulled ahead of the rest of us in the difficult task of inducing the fragile human cells to grow in culture, and he confirmed that they were indeed embryonic stem cells.

Far-Reaching Potential

In studies reported in the November 6, 1998, issue of *Science*, Thomson demonstrated that the human cells formed a wide variety of recognizable tissues when transplanted under the skin of mice. Discussing his results before an inquisitive subcommittee of the U.S. Senate, Thomson described how the cells gave rise to tissue like that lining the gut as well as to cartilage, bone, muscle and neural epithelium (precursor tissue of the nervous system), among other types. What is more, descendants of all three fundamental body layers of a mammalian embryo were represented. Some normally derive from the outermost layer (the ectoderm), others from the innermost or middle layers (the endoderm or mesoderm). This variety offered further evidence of the cells' developmental flexibility. Such results encourage hope that research on embryonic stem cells will ultimately lead to techniques for generating cells that can be employed in therapies for many conditions in which tissue is damaged.

If it were possible to control the differentiation of human embryonic stem cells in culture, the resulting cells could help repair damage caused by congestive heart failure, Parkinson's disease, diabetes and other afflictions. They could prove especially valuable for treating conditions affecting the heart and the islets of the pancreas, which retain few or no stem cells in an adult and so cannot renew themselves naturally. One recent finding hints that researchers might eventually learn how to modify stem cells that have partly differentiated so as to change the course of their development.

First, though, investigators will have to learn much more about how to induce embryonic stem cells to mature into desired tissues. Much of what is known so far has been gleaned from studies of mouse embryonic stem cells, which were the first to be characterized. Researchers derived them in 1981 from mouse embryos at the 100-cell stage. Such embryos consist of a hollow ball of cells known as a blastocyst. Hardly wider than an eyelash, a blastocyst has an internal thickening of its wall known as the inner cell mass. In a uterus, it would form the entire fetus and its membranes, such as the amnion.

When mouse blastocysts are cultured in a petri dish, the outer layer of cells soon collapses, and undifferentiated cells from the inner cell mass spontaneously form clumps that can be cultured to yield embryonic stem cells. These can grow and divide for long periods in an undifferentiated state. Yet when injected back into a mouse blastocyst, they respond to physiological cues, and mature cells derived from those stem cells appear in virtually the full range of the embryo's tissues. For this reason embryonic stem cells are termed pluripotent, from the Latin for "many capabilities." (Mouse embryonic stem cells are sometimes described as totipotent, implying that they can form all tissues, although they do not form placenta.) Embryonic stem cells thus have a lot in common with cells in the inner cell mass, the mothers of all cells in the body, but are not identical to them: subtle changes occur in culture that slightly limit their potential.

As investigators experimented with different culture conditions, they found that if a key biological chemical, known as leukemia inhibitory factor, is not supplied, the cells start differentiating in an unpredictable way. Interestingly, though, the repertoire of cell types that have arisen in this way is much smaller than that seen when the cells are injected into a blastocyst— probably because vital biological chemicals present in the embryo are not in the culture medium. This contrast raised the question of whether artificial conditions could be found that would mimic those in the embryo.

Directing Development

Such manipulations are possible. Gerard Bain and David I. Gottlieb and their associates at the Washington University School of Medicine have shown that treating mouse embryonic stem cells with the vitamin A derivative retinoic acid can stimulate them to produce neurons (nerve cells). That simple chemical seems to achieve this dramatic effect on the cells by activating a set of genes used only by neurons while inhibiting genes expressed in cells differentiating along other pathway.

My colleague Meri Firpo and her former coworkers in Gordon Keller's laboratory at the National Jewish Medical and Research Center in Denver had comparable success deriving blood cells. They discovered that specific growth factors stimulated cells derived from embryonic stem cells to produce the complete range of cells found in blood.

Embryonic stem cells might even generate some useful tissues without special treatment. I never cease to be amazed, when looking through a microscope at cultures derived from embryonic stem cells, to see spontaneously differentiating clumps beating with the rhythm of a heart. Investigators could potentially allow such transformations to occur and then select out, and propagate, the cell types they need.

Loren J. Field and his associates at the Indiana University School of Medicine have done just that. Employing a simple but elegant method, they enriched the yield of spontaneously differentiating heart muscle cells, or cardiomyocytes, to greater than 99 percent purity. To achieve that goal, they first introduced into mouse embryonic stem cells an antibiotic-resistance gene that had been engineered to express itself only in cardiomyocytes. After allowing the cells to differentiate and exposing them to enough antibiotic to kill cells that lacked the resistance gene, Field's team was able to recover essentially pure cardiomyocytes. Remarkably, when the cells were transplanted into the hearts of adult mice, the cardiomyocytes engrafted and remained viable for as long as seven weeks, the longest period the researchers analyzed.

Likewise, Terrence Deacon of Harvard Medical School and his

co-workers have transplanted embryonic stem cells into a particular region in the brains of adult mice. They observed that many of the engrafted cells assumed the typical shape of neurons. Some of those cells produced an enzyme that is needed to make the neurotransmitter dopamine and occurs in quantity in dopamine-secreting neurons. Others produced a chemical found in a different class of neurons. What is more, the nervelike cells in the grafts elaborated projections that resembled the long, signal-carrying neuronal branches known as axons; in the brain, some of these extended into the surrounding tissue. Whether such cells not only look normal but also function normally has not yet been assessed. Nor is it clear which (if any) growth factors in the mice stimulated the transplants to form neurons: surprisingly, nervelike cells also developed in grafts placed adjacent to the kidney.

The technique for establishing a culture of embryonic stem cells is more involved when primate embryos are the source, rather than mouse embryos. The outer cell layer of the primate blastocyst does not fall apart so readily in culture, so researchers must remove it, or the cells of the inner cell mass will die. But the results from the mouse studies suggest that as researchers gain experience with human embryonic stem cells, it will become possible to stimulate them to produce, at least, blood cells, heart muscle cells and neurons. Other medically valuable types might be achievable, such as pancreatic islet cells, for treatment of diabetes; skin fibroblasts, for treatment of burns or wounds; chondrocytes, for regenerating cartilage lost in arthritis; and endothelial (blood vessel-forming) cells, to repair blood vessels damaged by atherosclerosis.

Unfortunately, embryonic stem cells also have a dark side. The jumble of cell types they form when injected into mature mice constitutes a type of tumor known as a teratoma. Researchers will have to be sure, before using cells therapeutically, that they have all differentiated enough to be incapable of spreading inappropriately or forming unwanted tissue. Rigorous purification of such cells will be required to safeguard the recipients.

The cells that Gearhart obtained from developing ovaries and testes also show medical promise. They are called embryonic germ cells, because they are derived from the ancestors of sperm and eggs, which are together referred to as germ cells. Gearhart has shown that his cells, too, are pluripotent:

in the petri dish they can give rise to cells characteristic of each of the embryo's basic layers. As of this writing, however, Gearhart has not published details of what happens when embryonic germ cells are placed under the skin of mice, so information about their potential for tissue formation is still somewhat limited.

Challenges and Opportunities

All the differentiated cells discussed so far would probably be useful in medicine as isolated cells or as suspensions; they do not have to organize themselves into precisely structured, multicellular tissues to serve a valuable function in the body. That is good news, because organ formation is a complex, three-dimensional process. Organs generally result from interactions between embryonic tissues derived from two distinct sources. Lungs, for example, form when cells derived from the middle layer of the embryo interact with those of the embryonic foregut, which is derived from the inner layer. The process stimulates embryonic foregut cells to form branches that eventually become the lungs. For would-be tissue engineers, learning how to direct pluripotent stem cells through similar interactions with the goal of building entire organs will be hugely difficult. Nevertheless, some researchers are working on solutions to those very problems.

Another challenge is to create cells for transplantation that are not recognized as foreign by the recipient's immune system. This end could be achieved in principle by genetically altering human embryonic stem cells so they function as "universal donors" compatible with any recipient. Alternatively, embryonic stem cells genetically identical to the patient's cells could be created, as in the scenario of the heart attack victim described earlier.

The first option, creating a universal donor cell type, would involve disrupting or altering a substantial number of genes in cells. The changes would prevent the cells from displaying proteins on their outer surface that label them as foreign for the immune system. Yet bringing about this alteration could be hard, because it would require growing embryonic stem cells under harsh conditions, in particular exposing them to multiple rounds of selection with different drugs.

The second option, making cells that are genetically identical to the patient's tissues, involves combining embryonic stem cell technology and a fundamental step in cloning, as described in the vignette opening this article. Using a hollow glass needle one-tenth of the diameter of a human hair, a researcher would transfer a somatic (nonreproductive) cell—or just its gene-containing nucleus—into an unfertilized egg whose chromosomes have been removed. The egg would then be activated by an electrical shock, launching it on its developmental journey with only the genetic information of the transferred, or donor, cell.

In several animal studies on nuclear transfer, cells from existing adult animals have been used as the gene donors, and the altered cells have been implanted into the uterus of a living animal. These experiments gave rise to Dolly the sheep and to some mice and cattle as well. To create cells for transplantation with this combination of approaches, an investigator would use a cell from the patient as a donor but would culture the resulting embryo only until it reached the blastocyst stage. Then the embryo would be used to produce embryonic stem cells that were genetically identical to a patient's own cells.

Human embryonic stem cells could have other applications, too. Because the cells could generate human cells in basically unlimited amounts, they should be extremely useful in research efforts designed for discovering rare human proteins. These programs need great quantities of cells in order to produce identifiable amounts of normally scarce proteins. And because embryonic stem cells resemble cells in early embryos, they could be employed to flag drugs that might interfere with development and cause birth defects.

Finally, such cells offer an approach to studying the earliest events in human development at the cellular and molecular levels in a way that is ethically acceptable. The moral issues associated with experiments on embryos should not arise because embryonic stem cells lack the ability to form an embryo by themselves. Research on the cells could provide insights into fundamental questions that have puzzled embryologists for decades, such as how embryonic cells become different from one another, and what causes them to organize into organs and tissues. The lessons learned from mice, frogs, fish and fruit flies on these subjects are highly germane to humans. Yet understanding these processes in our own species will ultimately provide us with the greatest benefits and the deepest satisfaction.

Encapsulated Cells as Therapy

Michael J. Lysaght and Patrick Aebischer

In 1994 a man suffering from relentless pain became one of the first volunteers to test an entirely new approach to treating human disorders. As he lay still, a surgeon threaded a small plastic tube into his spinal column. The sealed tube, five centimeters long and as thin as the wire in a standard paper clip, contained calf cells able to secrete a cocktail of painkillers.

If all went well, the secretions would seep out the tube through minute pores and then diffuse inz the spinal cord. Meanwhile, nutrients and oxygen from the surrounding cerebrospinal fluid would slip into the capsule to sustain the cells. At the same time, the tubing would bar entry by large substances. Specifically, it would prevent cells and antibody molecules of the immune system (both of which are relatively big) from contacting the bovine cells and destroying them as foreign invaders.

The ultimate aim of this particular procedure is to relieve discomfort, by interrupting the flow of pain signals through the spinal cord to detection centers in the brain. The 1994 study, however, was preliminary. It was designed to see whether the implanted cells could survive and release their analgesics for months. They did. Similar success in several patients later justified a major trial, now under way to assess pain control directly.

But the results also had broader implications. They fueled growing optimism, based on extensive animal experiments, that combining living cells with protective synthetic membranes could help correct a range of human disorders.

Five years later, excitement over the strategy—variously known as encapsulated-cell, immunoisolation or biohybrid therapy—seemed entirely justified. Like the pain implant, a biohybrid live support system had progressed to a controlled human trial involving scores of patients and multiple centers.

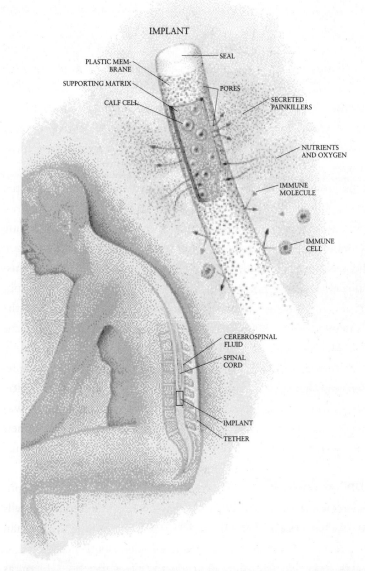

IMPLANT

PLASTIC MEM-
BRANE

SUPPORTING MATRIX

CALF CELL

SEAL

PORES

SECRETED
PAINKILLERS

NUTRIENTS
AND OXYGEN

IMMUNE
MOLECULE

IMMUNE
CELL

CEREBROSPINAL
FLUID

SPINAL
CORD

IMPLANT

TETHER

An implant under study for relief of chronic pain is fitted into a fluid-filled canal in the spinal column. It consists of a narrow plastic tube (detail), several centimeters long, filled with calf cells that secrete natural painkillers. Ideally, the painkillers will leak from the implant through tiny pores in the plastic, diffuse to nerve cells in the spinal cord and block pain signals from flowing to the brain. The pores will also allow small nutrients and oxygen to enter the implant but will be too small to permit access by immune components that normally destroy foreign cells. The tether allows the implant to be removed.

Most proposed applications involve implanting encapsulated cells in a selected site in the body. Some, though, such as the liver treatment, would incorporate cells and membranes into a bedside device resembling a kidney dialysis machine. Immunoisolation therapy appeals to us and other researchers because it overcomes important disadvantages of implanting free cells. Like free cells, those encased in membranes can potentially replace critical functions of ones that have been damaged or lost. They can also supply such "extras" as painkillers. They can even provide gene therapy, secreting proteins encoded by genes that molecular biologists have introduced into cells.

Free cells, however, are likely to be ambushed by the immune system unless they come from the recipients themselves or their twins. For that reason, patients usually require immune-suppressing drugs. By mechanically blocking immune attacks, plastic membranes around grafted cells should obviate the need for such medicines, which can predispose people to infection, certain cancers (lymphomas) and kidney failure.

The immune protection afforded by plastic membranes should also allow cells derived from animals to be transplanted into people. Unencapsulated animal cells are not a viable option, because existing immune-suppressing drugs do not fully protect against the rejection of cross-species implants (xenografts). Use of animal cells would help compensate for the well-known shortage of human donor tissue. Finally, cells implanted within a plastic casing can be retrieved readily if need be. Free cells, in contrast, often cannot be recovered.

An Inspired Proposal

Current efforts to encapsulate cells for therapy owe a great debt to ideas put forward by William L. Chick in the mid-1970s, when he was at the Joslin Research Laboratory in Boston. Like legions of scientists then and now, he had his sights set on curing insulin-dependent (type I) diabetes, which usually strikes youngsters. This disorder arises when the pancreas stops making insulin, a hormone it normally releases in amounts tuned to control the concentrations of glucose (a sugar) in the blood. Daily insulin injections save lives, but they do not mimic the natural pattern of insulin release from the pancreas. In consequence, tissues may at times become exposed to too much glucose. Over years, this excess can lead to such diabetic complications as blindness and kidney failure.

Chick thought implantations of encapsulated pancreatic islets—the clusters of cells that contain the insulin-secreting components—might restore the proper pattern of insulin release without requiring the administration of immunosuppressants. Use of islets from pigs (then the main source of injected insulin) would, moreover, guarantee a rich pipeline of cells.

Studies in rodents in the mid-1970s and thereafter suggested his logic was sound. Unfortunately, certain technical obstacles have so far kept immunoisolation therapy from fulfilling its promise in diabetes. Chick died in 1998, without seeing his vision fulfilled. His pioneering ideas have, nonetheless, sparked impressive progress on other fronts, including device design.

Creative Configurations

Encapsulation systems now come in a multitude of configurations. All, however, include the same basic ingredients: cells (typically ones able to secrete useful products), a matrix that cushions the cells and otherwise supports their survival and function, and a somewhat porous membrane. Biomedical engineers now know that cells in an implant will work poorly or die if they are farther than 500 microns (millionths of a meter) from blood vessels or other sources of nourishment—a distance roughly equivalent to the diameter of the graphite in a mechanical pencil.

Vascular, or flow-through, designs were the first to be tested (for correcting diabetes in rodents). These devices divert a patient's circulating blood into a plastic tube and then back to the circulatory system. Secretory cells are placed in a closed chamber that surrounds a slightly porous segment of the tubing, the way a doughnut surrounds its hole. As blood flows through this part of the circuit, it can absorb substances secreted by the therapeutic cells and can provide oxygen and nutrients to the cells. If islets are in the chamber, they will match the insulin released to the concentration of glucose in the blood. For other applications, cells that emit a product at a constant rate can be chosen.

Flow-through devices can be produced in implantable forms. But they will probably find most application in bedside equipment, because implants require invasive, vascular surgery and long-term administration of blood thinners (to prevent blood clots from forming in the tubing). In addition, if an implanted tube breaks, internal bleeding will result.

Searching for a less invasive method, researchers introduced "microencapsulation" in the late 1970s. To make microcapsules, workers put a single pancreatic islet or a few thousand individual cells into a drop of aqueous solution containing slightly charged polymers. Then they bathe the drop in a solution of oppositely charged polymers. The polymers react to form a coating around a cell-and-fluid-filled droplet measuring about 500 microns in diameter.

Microcapsules are easy to produce and thus are valuable for quick experiments but have notable drawbacks for human therapy. They are quite fragile. Once placed, they may be difficult to find and remove—a distinct problem if they have unwanted effects. What is more, the volume needed to correct a disorder may be too great to fit conveniently in a desired implant site.

The most practical format for human therapy appears to be performed macrocapsules, initially empty units that are loaded with a matrix and all the cells needed for treatment. Some macrocapsules are disks about the size and shape of a dime or a quarter. Others are roughly the size and shape of the stay in a shirt collar. Usually, however, macrocapsules intended for humans take the form of a sealed tube, or capillary, that is several centimeters long and between 500 and 1,000 microns in diameter.

Macrocapsules are far more durable and rugged than microcapsule droplets, contain internal reinforcements, can be tested for seal integrity before implantation and can be designed to be refillable in the body. They can also be retrieved simply. Their main limitation is the number of cells they can accommodate: up to about five million for a tube and up to 50 or 100 million for a disk or flat sheet. Those figures are adequate for many applications, but not all. Enlarging the capsules can render them prone to bending, which promotes breakage. In addition, the edges of bent regions encourage fibrosis, an ingrowth of local tissue. Fibrosis can choke off transport to and from encapsulated cells.

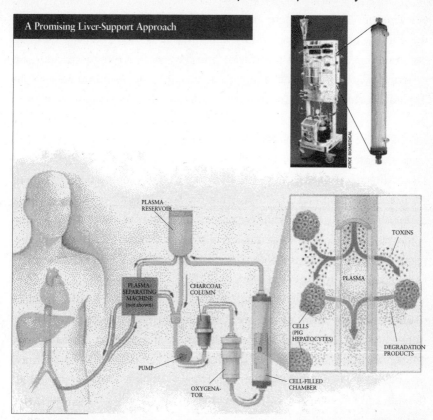

A Promising Liver-Support Approach

CIRCE BIOMEDICAL

PLASMA
RESERVOIR

PLASMA
SEPARATING
MACHINE
(not shown)

CHARCOAL
COLUMN

PUMP

OXYGENA-
TOR

CELL-FILLED
CHAMBER

TOXINS

PLASMA

CELLS
(PIG
HEPATOCYTES)

DEGRADATION
PRODUCTS

Not all encapsulation systems are implants. Liver-support systems currently being studied operate outside the body. They aim to sustain liver-failure patients until a compatible organ becomes available for transplantation. The particular device shown and illustrated above was developed by teams led by Claudy J. P. Mullon of Circe Biomedical in Lexington, Massachusetts, and Achilles A. Demetriou of Cedars-Sinai Medical Center in Los Angeles.

This machine draws blood from a patient and pumps the fluid component (plasma) through a charcoal column (meant to remove some toxins) and an oxygen-replenishing unit before delivering it to a chamber containing healthy liver cells—hepatocytes—from pigs. In the chamber (detail), the plasma courses through slightly porous tubes, which are surrounded by the hepatocytes. Toxins from the plasma diffuse into the cells, which are intended to convert the poisons into innocuous substances. After the purified plasma leaves the chamber, it recombines with blood cells and is returned to the patient.

Fabricators of shunts, microcapsules and macrocapsules aim for a membrane pore size that will allow diffusion of molecules measuring up to 50,000 daltons, or units of molecular weight. Holes that size generally are small enough to block invasion by immune cells and most immune molecules but are large enough to allow the inflow of nutrients and oxygen and the outflow of proteins secreted by implanted cells. Actual membranes end up containing a range of pore sizes, however, and so some large immune system molecules will inevitably pass through the membranes into an implant. Fortunately, this phenomenon does not undermine most implants.

New Focus on Designer Cells

Until the late 1980s, most biohybrid devices relied on primary cells: those taken directly from donor tissue. Primary cells are convenient for small studies in small animals, but obtaining the large quantities needed for big animals (including humans) or for numerous recipients can be problematic. And because every donor has its own history, guaranteeing the safety of primary cells can be a formidable undertaking. In the early 1990s, therefore, some teams began turning to cell lines.

These lines consist of immortal, or endlessly dividing, cells that multiply readily in culture without losing their ability to perform specialized functions, such as secreting helpful substances. Many primary cells replicate poorly in culture or have other disadvantages. Hence, to make a cell line, investigators often have to alter the original versions. Once established, though, cell lines can provide an ongoing supply of uniform cells for transplantation.

The potential utility of cell lines for encapsulation therapy became abundantly clear in animal tests that we and our colleagues performed starting in 1991. The well-established PC-12 line, derived from a rodent adrenal tumor (a pheochromocytoma), was known to secrete high levels of dopamine, a signaling molecule depleted in the brains of patients with Parkinson's disease. To see whether implants containing these cells might be worth studying as a therapy for Parkinson's, we put small tubes of the cells into the brains of diverse animals whose dopamine-producing cells had been chemically damaged to produce Parkinson's-like symptoms. In many subjects, including nonhuman primates, the procedure dramatically reversed the symptoms.

Significantly, the cells did not proliferate uncontrollably and puncture the capsules. They replaced cells that had died but did not allow the population to exceed the carrying capacity of the implant. The studies also eased fears that if immortalized cells escaped, they would inevitably spawn cancerous tumors. Immortalization is one step on a cell's road to cancer. To be truly malignant, though, cells must acquire the ability to invade neighboring tissue, grow their own blood supply and spread to distant sites. Tumor formation is a potential concern in same-species transplants of immortalized cells, but cross-species transplants turn out to be less worrisome: unencapsulated rat PC-12 cells in primate brains did not generate tumors. In fact, they did not even survive; the recipients' immune systems destroyed them quickly.

The PC-12 work was never followed up in Parkinson's patients, perhaps because other promising treatments took precedence. Still, the studies did demonstrate the feasibility of deploying cell lines in immunoisolation therapies.

Success with cell lines also opened the door to use of genetically modified cells, because dividing cells are most amenable to taking up introduced genes and producing the encoded proteins. In other words, immunoisolation technology suddenly offered a new way to provide gene therapy. Molecular biologists would insert genes for medically useful proteins into cell lines able to manufacture the proteins, and the cells would then be incorporated into plastic-covered implants.

Gene therapy protocols frequently remove cells from patients, insert selected genes, allow the altered cells to multiply and then return the resulting collection to the body in the hope that the encoded proteins will be made in the needed quantities. The output of capsules filled with genetically altered cells, in contrast, can be measured before the implants are delivered to patients. Later, the capsules can be removed easily if need be.

An unresolved issue is whether cell lines enlisted for encapsulation therapies should be derived from animals or humans. Primary cells, taken directly from donors, almost certainly need to come from animals, because human donors tissue is in such short supply. Some researchers prefer animal-derived lines because renegade cells that broke free from an implant, being highly foreign to the recipient, would meet the promptest immune destruction. To provide human therapeutic proteins to a recipient, cell designers could readily equip the animal cells with human genes for those proteins. Other workers favor human-derived lines, in part because human cells tend to fare better within capsules. They also skirt the risk that animal pathogens will be transferred to humans. For extra safety, human cells could be engineered to elicit swift immune recognition should they escape.

Readers may notice that encapsulated cells, genetically altered or not, often essentially serve as delivery vehicles for therapeutic proteins. Yet proteins can be delivered by injection. Why, then, would cell implantation be needed at all?

Encapsulated-cell therapy can be very helpful when injections cannot provide enough of a protein where it is needed most, such as in a tumor or behind the blood-brain barrier—a natural filter that blocks many blood-borne substances from reaching cells of the brain and spinal cord. Encapsulated cells would likewise be valuable when a therapeutic protein is too unstable to be formulated as a drug or when reproducing a natural pattern of protein delivery is important (as with diabetes).

Human Studies

Genetically manipulated cell lines are likely to predominate in biohybrid devices of the future. Yet applications involving primary cells, having been studied the longest, have progressed to the most definitive clinical tests. One example, developed by the two of us and a large contingent of associates, is the treatment for chronic pain described at the start of this article. We obtain the cells for the pain implant from the adrenal glands of calves raised under highly controlled conditions. Certain adrenal components—the chromaffin cells—naturally release a suite of analgesics. After carefully purifying about three million of these cells, we fit them into a hollow fiber that is sealed at both ends, linked to a tether (for retrieval) and implanted, through a minimally invasive procedure, into the spinal column.

When our surgical collaborators showed in the mid-1990s that such implants could function for months in patients, they noted possible signs of pain control. Several patients reported significant reductions in discomfort and in morphine use. But these experiments did not include a comparison group receiving a placebo (say, an empty capsule), so we could not be sure whether our treatment was truly responsible. The large clinical trial now in progress involves more than 100 patients and is designed specifically to address the extent of pain relief. It is headed by Moses B. Goddard of CytoTherapeutics in Lincoln, Rhode Island.

Regardless of the outcome, the data already in hand demonstrate that immunoisolated cells of animal origin can live for months in the central nervous system of subjects who are taking no immune-suppressing drugs. In contrast, no organ transplanted from animals to humans has survived without encapsulation even when supported by the aggressive delivery of immuno-suppressants.

A second well-developed application of immunoisolation therapy—the liver-assist device—also relies on cells harvested directly from animals. Within a healthy liver, cells known as hepatocytes take up toxins and break them into innocuous forms. When the liver fails, such toxins can accumulate to lethal levels. Liver transplants can save patients, but many die waiting for a donor organ matched to their tissue type. The biohybrid liver systems under study aim to keep patients alive until a donor is found.

This "bridge to transplant" therapy employs a bedside, vascular apparatus. In essence, blood from the patient is pumped to a closed chamber in which a semiporous segment of the blood-carrying tube is surrounded by a suspension of pig hepatocytes. The hepatocytes take up the toxins from the flowing blood and degrade them, so that healthier blood returns to the body as it completes its circuit. In contrast to the pain implant, which delivers a few milligrams of cells and is expected to function continuously for months or years, a liver device can accommodate between 20 and 200 grams of purified hepatocytes (about the weight of the meat in a Big Mac) and would be needed for just six to 24 hours at a time.

In an initial study involving close to 40 patients with terminal liver failure, the equipment tested functioned exactly as hoped. That finding, reported in 1998, paved the way for a large, controlled trial now starting in the U.S. and Europe. Expectations of success are high, and researchers have reason to believe that under special circumstances, such as acute liver failure caused by excessive intake of acetaminophen, the liver might regenerate and that no transplantation would be required. Yet optimism has to be tempered by past experience. The path to liver-support systems is littered with interventions that worked beautifully in initial tests but faded in large trials.

For both the pain and liver applications, scientists must address concerns that genes from unknown animal viruses might be hiding in the harvested cells and that those genes might give rise to viruses and to dangerous infections in transplant recipients. (Rigorous screening methods ensure that no already known pathogen is transferred.) Fortunately, plastic membranes should provide a formidable barrier to the transmission of animal viruses, and to date, no patient has acquired even a benign infection from donor cells. Even so, investigators are neither cavalier nor complacent about the issue and continue to attend to it closely.

Although less advanced, human trials of gene therapy applications have begun as well. Two small studies target disorders of the central nervous system. The first clinical trial of genetically altered encapsulated cells took aim at amyotrophic lateral sclerosis (ALS), the neurodegenerative condition marked by decay of spinal nerves that control the muscles. ALS killed baseball legend Lou Gehrig. In 1996 six patients received implants containing a cell line—derived from baby hamster kidney cells—that had been given the gene for a protein called ciliary-derived neurotrophic factor (CNTF). This gene was chosen because other studies in animals and people had suggested the factor might retard the deterioration of neurons that usually die in ALS patients. The protocol was much like that used for chronic pain: a small tube filled with cells was implanted in the spinal column.

The study examined whether the cells survived and released potentially therapeutic amounts of CNTF throughout the three-month experiment. The cells functioned well. The treatment did not appear to retard disease progression, however, although the test had too few subjects and was too short to be particularly informative on that score. Nevertheless, this trial suggested that if the right gene or mix of genes for treating ALS were found, encapsulated cells would serve as a good means of delivering them to the central nervous system.

Implants containing the same cell line are now being evaluated in patients with Huntington's disease, which progressively kills certain brain cells. This time, however, the capsules have been placed into fluid-filled spaces in the brain called ventricles. This gene therapy protocol is being conducted in Paris and has just started. A number of animal experiments evaluating immunoisolation for the delivery of gene therapy have begun as well.

The Special Challenge of Diabetes

If immunoisolation research is progressing well in many areas, why has no one managed, after more than 20 years of trying, to perfect an islet-cell encapsulation procedure to correct diabetes?

After 1977, when Chick and his colleagues reversed diabetes in rodents, at least a dozen laboratories around the world replicated that feat, with a wide range of implant designs in various rodent models of diabetes. But immunoisolation therapy based on islets has not fared well in larger species, such as dogs, monkeys and humans. The most positive results come from single cases. Moreover, on close examination, many of those reported successes have been achieved only with the help of immunosuppressants or some amount of injected insulin.

Much of the difficulty stems from the sheer number of islets required by large animals and humans: 700,000 or so, sheltering approximately two billion insulin-producing, "beta" cells. That amount is nearly 1,000 times greater than the cell volume encapsulated successfully in clinical implants to date. Diabetes in mice can be reversed with only about 500 islets, which technicians generally extract from donor pancreases by hand. But hand-picking 700,000 islets is out of the question, and semiautomated techniques have not been able to isolate the required quantities of healthy islets consistently. Further, in the native pancreas, each islet enjoys its own blood supply. Islets suffer in the spartan environment within implanted capsules. For these reasons and others, we agree with those who have concluded that an implanted semiartificial pancreas based on encapsulated islets is most likely to remain an unfulfilled goal for the foreseeable future.

A new proposal, though, just might break the impasse. Employing a variety of approaches, at least three groups are developing cell lines that release insulin in response to the same complex signals that trigger insulin secretion from the healthy pancreas. The plan is to create cells that produce more insulin than natural beta cells (so that fewer cells would be required) and that are equipped to survive in the nutrient-poor, oxygen-depleted environment of an implant. Five years ago creation of such cells might have been regarded as impossible, but recent progress in cell and molecular biology has been overwhelming.

We expect to see glucose-responsive, insulin-secreting cell lines tested in large animals five years from now, possibly much sooner. And we are hopeful that those lines will progress rapidly to the clinic after that. Some experts believe this prediction is too conservative; others counsel that the goal will take longer to achieve. But everyone agrees that an artificial pancreas or a biohybrid version must continue to rank as a top priority for twenty-first-century medicine. As that work continues, new applications for immunoisolation therapy should arise as well. Indeed, we anticipate that over the next 20 years, encapsulated-cell therapy will emerge from its investigative stages to play a key role in treating some of the most refractory and debilitating diseases of humankind.

Couture Cures: This Drug's for You

Karen Hopkin

"One pill makes you larger and one pill makes you small. And ones that Mother gives you don't do anything at all." Some things were so simple in the '60s. If Grace Slick were to sing of today's pharmacology, her verse would probably sound more like the fine print at the bottom of a glossy drug ad: This pill may make you larger or smaller. It may also cause headaches, vomiting, night blindness, impotence and heart failure.

Of course, pharmaceutical companies want to avoid litigation when they market their medications to the public. But the long list of possible effects—and side effects—that accompanies every drug on the market today also reflects the recognition that individuals differ in the way they respond to medications. And that response depends, in large part, on a person's genes.

Now scientists are beginning to take advantage of new techniques that allow them to collect and compare large volumes of information about gene sequences—and about drug action—to predict how a person will respond to a given drug. These techniques stand to speed up the way drugs are designed and tested and may even change the way doctors diagnose and treat disease in the future.

Researchers have long known that genetic alterations can lead to disease. Mutations in one gene cause cystic fibrosis; in another gene, sickle cell anemia. But it is now becoming clear that genetic differences can also affect how well a person absorbs, breaks down and responds to various drugs. The cholesterol-lowering drug pravastatin, for example, does nothing for people with high cholesterol who have a common variant of an enzyme called cholesterol transfer protein.

Genetic variations can also render drugs toxic to certain individuals. Isoniazid, a tuberculosis drug, causes tingling, pain and weakness in the limbs of those who are termed slow acetylators. These individuals possess a less active form of the enzyme N-acetyltransferase, which normally helps to clear the drug from the body. Thus, the drug can outlive its usefulness and may stick around long enough to get in the way of other, normal biochemical processes. If slow acetylators receive procainamide, a drug commonly given after a heart attack; they stand a good chance of developing an autoimmune disease resembling lupus.

Balm or Bane?

Enter pharmacogenomics, a new science that aims to use a systematic genome-wide analysis of genetic variation to see which drugs might work for you and which might make you sicker. The clues come in the form of single nucleotide polymorphisms, or SNPs (pronounced "snips")—genetic hot spots scattered along our chromosomes that can vary in DNA sequence from person to person. Researchers are now compiling an extensive catalogue of these SNPs in the hope that they will be able to link particular genetic fingerprints with differences in drug response.

SNP testing would work something like this: a doctor or technician would extract DNA from a small sample of a person's blood or other body cells. The DNA would then be washed over a SNP chip—a glass slide studded with DNA fragments that represent all the common genetic variations in, say, a gene known to control how well a drug is absorbed. (Some SNPs correlate with good absorption and some with poor absorption.) The DNA from the patient would stick to whichever SNP it matched, and a scanner could then look at the chip and determine whether the person would be able to absorb the drug in question.

Drug vending machines that dole out designer doses on demand probably won't be popping up on street corners anytime soon. But scientists envision a day when physicians will prescribe pharmaceuticals tailored to our own specific genetic information, which we might carry around encoded on a credit-card-size plastic plate.

But beyond improving diagnostics, drug companies hope that pharmacogenomics will help them get more novel drugs to market. Currently 80 percent of drugs are shot down in early clinical trials because they are not effective or are even toxic, according to the Tufts Center for the Study of Drug Development at Tufts University. Pharmaceutical companies would like to boost the success rate of drug approval by testing new drugs only in individuals who are likely to show benefits from them during the clinical trial.

The problem is that people who are deemed genetically unresponsive might then fall through the cracks, observes William A. Haseltine, CEO of Human Genome Sciences in Rockville, Maryland. As it stands, pharmacogenomics is headed toward splintering the drug market, generating three or four different drugs that each might treat only tens of thousands of individuals with a particular disease—a scenario Haseltine views as "utter folly." Instead he favors using pharmacogenomics to develop new drugs aimed at treating the majority of people.

Using pharmacogenomics to select people who will respond to new drugs, Haseltine notes, "is a route around, not through, a major problem"—the problem being that it is difficult to develop drugs that work. Indeed, many companies are pursuing different methods for stepping up the flow through the pharmaceutical development pipeline. The goal, simply put, is to be able to generate and test the largest number of compounds in the shortest amount of time with the least amount of human effort. So researchers are turning to robots that can simultaneously analyze tiny volumes of thousands of samples—a process dubbed high-throughput screening. Then they use computers to process and keep track of all the results—and, in some cases, to suggest which drugs should be tested.

Researchers at Neurogen, a pharmaceutical company in Branford, Connecticut, for example, use high-throughput computer modeling methods to select the most promising drugs from a "virtual library," a computer database that contains the molecular structures of billions and billions of chemical compounds not yet made. Say they want to develop a more effective antianxiety medication. The scientists browse through a few hundred million molecules in their virtual library and select a few dozen groups of compounds that might interact with the particular types of satellite-dish-like proteins called receptors on the surfaces of nerve cells in the brain that are specifically associated with anxiety. Drugs that bind to these receptors could prevent panic attacks by interfering with the chemistry that makes some people unnecessarily anxious. The compounds could then be synthesized and tested, and the results could be used to home in on the most promising antianxiety drugs. Combining such rational drug design with powerful computing tools allows investigators to test thousands of compounds in a matter of weeks, says Neurogen's vice president Charles Manly.

But pharmaceutical companies are seeking to do more than just increase the number of drugs they test: they are also looking for better ways to select the best drugs early in the process. One way they are doing this is by making early drug screening richer in information. Instead of just testing whether a compound can bind to a receptor, for instance, researchers are developing high-throughput assays to measure how strong the binding is and how the drug affects the various biochemical processes of a cell. Does it switch on the correct genes and proteins, for example, or does it shut them off? Testing a drug's selectivity, toxicity, metabolism and absorption at the start of the screening process will cut down on efforts wasted on trying ineffective drugs in humans.

Living Chips

Eventually, scientists will be able to assay compounds on living cells that are growing on silicon chips, says D. Lansing Taylor of Cellomics in Pittsburgh. He and his colleagues are now developing such a cell chip for detecting agents of biological warfare. The device, dubbed a "canary on a chip," is a prepackaged piece of silicon covered with living nerve cells from insects. Many of the bacteria believed to be favored by bioterrorists secrete nerve toxins, so these chips could provide an early warning of a biological attack.

Such cell-chip technology might also allow doctors to determine which kinds of chemotherapies would work best for a cancer patient. A physician could biopsy a tumor, grow the harvested cells on a chip and then test to see which chemicals would be most effective at killing the cells. Testing the cells themselves could save the patient from undergoing a series of unnecessary and ineffective treatments.

For some of these technologies, the future is already here. Affymetrix in Santa Clara, California, now offers a SNP chip that can be used to detect 18 variants of the gene that codes for cytochrome P450—a liver enzyme responsible for breaking down nearly one-quarter of all commonly prescribed drugs. The company should soon release HuSNP, a DNA chip that will allow researchers or physicians to characterize genetic variations at 1,500 different marker sequences, which will help them link individual variations to different diseases. And in the next few years workers at the National Institutes of Health's National Human Genome Research Institute (NHGRI)—and at the 10 pharmaceutical companies that recently banded with the Welcome Trust to form the SNP Consortium—expect to generate a map containing some 400,000 SNPs.

And that's when the fun will begin. "We'll have this catalogue of SNPs, but we'll still have to figure out which ones are associated with disease risk or drug response," says Francis S. Collins, director of the NHGRI. Then disease by disease, drug by drug, investigators will need to compare thousands of individuals—people who respond well to a drug and those who respond poorly, for example—and determine how they differ at every one of these 400,000 SNPs. "That's a lot of SNPs," Collins notes. But the potential benefits—to drug companies and to society—are sure to be greater than the considerable challenge.

Personal Pills

Even before the human genome is fully decoded, academic and industry researchers have begun to take the next step: comparing how genetic information varies from individual to individual. The databases compiled from these endeavors will provide a record of human migrations and will show how multiple genes contribute to common diseases. But biotechnology and pharmaceutical companies also want to use this knowledge to tailor drugs to certain groups of patients. A customized pharmaceutical might eliminate life-threatening adverse reactions. And knowing how genetically distinct individuals react differently to a certain compound may reduce the cost of clinical trials by targeting only those patients capable of responding to a drug. Pharmacogenomics is the term that has evolved to describe the use of advanced genetic tools to elucidate how variations in patients' DNA may diminish or amplify drug effects or render a pharmaceutical toxic. Earlier this year an article in the *Journal of the American Medical Association* estimated that adverse drug reactions accounted for more than two million hospitalizations and more than 100,000 deaths in 1994, making them a leading cause of mortality in the U.S.

Many of the ideas that underlie pharmacogenomics are not new. It has been understood for decades that genes affect the way patients respond to drugs. For instance, pharmaceutical researchers sometimes look at how differences in the genes for liver enzymes called cytochrome P450 affect how patients metabolize a new drug candidate. But until now the genes one could study for such variations were few in number. The tools for rapidly compiling large compendiums of the minute variations in nucleotides (DNA bases) are of recent vintage.

Indeed, a race is under way to catalogue genetic variations among these single DNA bases, known as single nucleotide polymorphisms (SNPs, pronounced "snips"), which can be used in characterizing drug responses. The National Institutes of Health has launched a $36 million, three-year program to collect data on 50,000 to 100,000 SNPs, a new goal for its Human Genome Project. The information would be used not only to gauge drug responses but also to study disease susceptibility and to conduct basic research on population genetics. In midsummer a group of pharmaceutical companies discussed forming a consortium with the NIH that would supply additional funding and research resources to create an even larger public database. One impetus to establish a tie with industry has been a concern that private attempts to patent SNPs could choke off access to data for basic research. "These research tools are far upstream of any particular product," notes Francis S. Collins, who oversees the Human Genome Project at the NIH. "The public is best served by having them accessible to any researcher who wants to use them."

A pharmaceutical industry collaboration with the NIH would promote public access to SNPs. Still, some biotechnology companies have rushed to embrace pharmacogenomics by creating private databases. A French company, Genset, is testing the DNA of more than 100 people to develop a map of the entire human genome. The Genset map will contain 60,000 SNPs that are within or near genes that cause disease or differing drug reactions. Genset's chief genomics officer, Daniel Cohen, devised the first rough physical map of the human genome in 1993.

Abbott Laboratories, a major U.S. pharmaceutical manufacturer, has invested $20 million in Genset. The companies will market SNP map data to drug companies that wish to pinpoint during clinical trials a common set of variant nucleotides shared by people who do not respond to a drug. This information could then be used to create diagnostic tests to filter out unresponsive patients. Abbott, in fact, is paying Genset an additional $22.5 million to help it develop a diagnostic test to screen patients for zileuton, its own asthma drug, which can induce liver toxicity in 3 percent of patients. Genset is not the only one putting together SNP databases. In August, Incyte Pharmaceuticals announced plans to purchase Hexagen in Cambridge, England, as part of its effort to detect genetic variation.

The application of rapid tools for screening SNPs may eventually make it possible to look for the unique signature of an individual's DNA in a matter of hours. Traditional gene-sequencing technology might take two weeks and $20,000 to screen a single patient for variations in 100,000 SNPs. "That's going to make this prohibitive to put into a clinical-trial kind of system," noted Robert Lipshutz of DNA chipmaker Affymetrix at the annual meeting of BIO, a biotechnology industry trade group. Affymetrix is testing a chip that can detect 3,000 SNPs in less than 10 minutes. As the technology progresses, Affymetrix expects to be able to mill through 100,000 SNPs dispersed through a patient's genome in several hours, for as little as a few hundred dollars.

Not everyone wants to assess patient drug response by scanning the entire genome. Variagenics in Cambridge, Massachusetts, selects a few target genes thought to be associated with drug responses for a given disease, a more established approach intended to speed assessment of drug safety and refinement of diagnostic tests. "Genes involved in drug action are overrepresented among the genes whose sequences are already known," says Fred D. Ledley, the company's chief executive. To locate SNPs, an enzyme called resolvase scans the selected genes. It cuts the DNA when it finds a nucleotide that differs from a reference sequence. Using these data allows investigators to glean the genetic profile of patients who experience ill effects from a drug. One of Variagenics's goals is to improve the prescription of existing drugs. It is fashioning a test that will let physicians adjust the dosage of a widely prescribed cancer drug, 5-fluorouracil, that produces severe gastrointestinal side effects in some patients.

Before genetic profiling for drug prescriptions becomes routine, pharmacogenomics must overcome other obstacles. Individualizing pharmaceuticals may not necessarily sit well with big pharmaceutical companies, which are constantly in search of blockbuster drugs to offset multimillion-dollar development costs. A drug tailored to a specific subpopulation may fragment and diminish markets. Several drugs may be needed for a given condition, one for each genetic subtype. This strategy might still work if a manufacturer can charge enough for each drug. The real push toward pharmacogenomics may be driven by managed health care. A diagnostic test, even if it does add cost, could avoid the expense of today's trial-and-error methods of making multiple doctor's visits to have a prescription adjusted.

The hazards of placing patients in subgroups has not gone unnoticed. Without safeguards, health insurance providers might deny coverage to those with a certain genetic profile—patients for whom a drug is too expensive or for whom there is no treatment. The Human Genome Project's Ethical, Legal and Social Implications program will make the use of information about genetic variation its "number one priority" during the next five years, Collins says. "When you're cataloguing large numbers of SNPs on large numbers of people, it greatly accelerates the potential for this information to be misused in discriminatory ways," he remarks.

And according to one biotechnology industry leader, pharmacogenomics may simply be an ill-chosen approach to designing new drugs. William A. Haseltine, chairman and chief executive of Human Genome Sciences, asserts that pharmaceutical companies should be using genetic technologies to find the safest possible drug, not trying to save failed candidates by targeting them to selected patients. Diagnostic tests can be unreliable, he notes, and some patients could still sustain life-threatening reactions. Moreover, the multiple genes involved in a drug reaction can be hard to decipher. Environmental factors—food, other drugs ingested, a patient's gender and overall state of health—may account for much of how someone responds to a drug. "You've got to consider the whole person when using a drug," Haseltine says. "The pharmacogenomic argument is very similar to the sociobiology argument that everything is in the genes, when it is not." Debate may never fully settle the issue. Technology that can identify a patient's distinctive genetic profile—and thus alter the way drugs are prescribed—may always prove contentious.

—*Gary Stix, editor,* Scientific American

Suggested Reading

Aebischer, P., and M. J. Lysaght. "Immunoisolation and Cellular Xeno-transplantation," *Xeno* 3, no. 3 (June 1995): 43–48. Also available at http://www.ribotech.com/xeno on the World Wide Web.

Deacon, T. et al. "Blastula-Stage Stem Cells Can Differentiate into Dopaminergic and Serotonergic Neurons after Transplantation," *Experimental Neurology* 149 (January 1998): 28–41.

Emerich, D. F. et al. "Treatment of Central Nervous System Diseases with Polymer-Encapsulated Xenogenic Cells." In *Cell Transplantation for Neurological Disorders*. Thomas B. Freeman and Hakan Widner, ed. New York: Human Press, 1998.

Kennedy, M. et al. "A Common Precursor for Primitive Erythropioesis and Definintive Haematopoiesis," *Nature* 386 (April 1998): 488–493.

Klug, Michael G. et al. "Genetically Selected Cardiomyocytes from Differentiating Embryonic Stem Cells Form Stable Intracardiac Grafts," *Journal of Clinical Investigation* 98, no. 1 (July 1996): 216–224.

Lanza, Robert P., Robert Langer, and William L. Chick. *Principles of Tissue Engineering*. New York: R. G. Landes Company, 1997.

Lanza, Robert P., David K. C. Cooper, and William L. Chick. "Xenotransplantation," *Scientific American* 227, no. 1 (July 1997): 54–59.

Pedersen, Roger A. "Studies of in Vitro Differentiation with Embryonic Stem Cells," *Reproduction, Fertility and Development* 6, no. 5 (1994): 543–552.

Thomson, J. A. et al. "Embryonic Stem Cells Lines Derived from Human Blastocysts," *Science* 282 (November 6, 1998): 1145–1147.

Chapter 4

Working with the Genome

The Human Genome Business Today

Kathryn Brown

By the time you read this, you'll also be able to read the entire genetic code of a human being over the Internet. It's not exactly light reading—start to finish, it's nothing but the letters *A, T, C* and *G*, repeated over and over in varying order, long enough to fill more than 200 telephone books. For biologists, though, this code is a runaway bestseller. The letters stand for the DNA chemicals that make up all your genes, influencing the way you walk, talk, think and sleep. "We're talking about reading your own instruction book," marvels Francis S. Collins, director of the National Human Genome Research Institute in Bethesda, Maryland. "What could be more compelling than that?"

Collins heads the Human Genome Project (HGP), so far a $250 million effort to write out the map of all our genes. The HGP is a publicly funded consortium that includes four large sequencing centers in the U.S., as well as the Sanger Center near Cambridge, England, and labs in Japan, France, Germany and China. Working together for more than a decade, over 1,100 scientists have crafted a map of the three billion DNA base pairs, or units, that make up the human genome. And they are not alone. In April 2000 a brash young company called Celera Genomics in Rockville, Maryland, beat the public consortium to the punch, announcing its own rough draft of the human genome. The rivalry has cast a spotlight on the human genetic code—and what, exactly, researchers now plan to do with it.

"For a long time, there was a big misconception that when the DNA sequencing was done, we'd have total enlightenment about who we are, why we get sick and why we get old," remarks geneticist Richard K. Wilson of Washington University, one partner in the public consortium. "Well, total enlightenment is decades away."

But scientists can now imagine what that day looks like. Drug companies, for instance, are collecting the genetic know-how to make medicines tailored to specific genes—an effort called pharmacogenomics. In the years to come, your pharmacist may hand you one version of a blood pressure drug, based on your unique genetic profile, while the guy in line behind you gets a different version of the same medicine. Other companies are already cranking out blood tests that reveal telltale disease-gene mutations—and forecast your chances of coming down with conditions such as Huntington's disease. And some scientists still hold out hope for gene therapy: directly adding healthy genes to a patient's body. "Knowing the genome will change the way drug trials are done and kick off a whole new era of individualized medicine," predicts J. Craig Venter, president of Celera.

Even with the human code in hand, however, the genomics industry faces challenges. Some are technical: it's one thing to know a gene's chemical structure, for instance, but quite another to understand its actual function. Other challenges are legal: How much must you know about a gene in order to patent it? And finally, many dilemmas are social: Do you really want to be diagnosed with a disease that can't be treated—and won't affect you for another 20 years? As scientists begin unraveling the genome, the endeavor may come to seem increasingly, well, human.

The "Race"

All eyes were on the first finish line in the genome: a rough-draft sequence of the 100,000 or so genes inside us all. The HGP's approach has been described as painstaking and precise. Beginning with blood and sperm cells, the team separated out the 23 pairs of chromosomes that hold human genes. Scientists then clipped bits of DNA from every chromosome, identified the sequence of DNA bases in each bit, and, finally, matched each snippet up to the DNA on either side of it in the chromosome. And on they went, gradually crafting the sequences for individual gene segments, complete genes, whole chromosomes and, eventually, the entire genome. Wilson compares this approach to taking out one page of an encyclopedia at a time, ripping it up and putting it together again.

In contrast, Celera took a shorter route: shredding the encyclopedia all

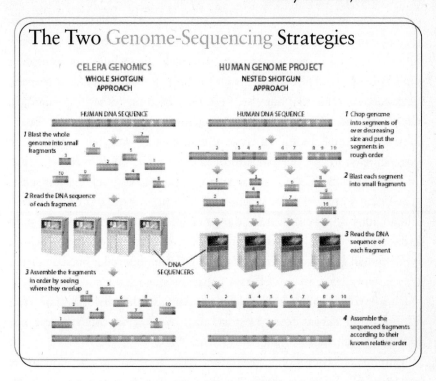

The Two Genome-Sequencing Strategies

CELERA GENOMICS
WHOLE SHOTGUN
APPROACH

HUMAN GENOME PROJECT
NESTED SHOTGUN
APPROACH

at once. Celera's so-called shotgun sequencing strategy tears all the genes into fragments simultaneously and then relies on computers to build the fragments into a whole genome. "The emphasis is on computational power, using algorithms to sequence the data," says J. Paul Gilman, Celera's director of policy planning. "The advantage is efficiency and speed."

The HGP and Celera teams disagree over what makes a "finished genome." Last spring Celera announced that it had finished sequencing the rough-draft genome of one anonymous person and that it would sort the data into a map in just six weeks. But the public team immediately cried foul, as Collins noted that Celera fell far short of its original genome-sequencing goals. In 1998, when the company began, Celera scientists planned to sequence the full genomes of several people, checking its "consensus" genome 10 times over. In its April announcement, however, Celera declared that its rough genome sequencing was complete with just one person's genome, sequenced only three times.

Although many news accounts have characterized the HGP and Celera as competing in a race, the company has had a decided advantage. Because the HGP is a public project, the team routinely dumps all its genome data into GenBank, a public database available through the Internet (at www.ncbi. nlm.nih.gov/). Like everyone else, Celera has used that data—in its case, to help check and fill the gaps in the company's rough-draft genome. Essentially Celera used the public genome data to stay one step ahead in the sequencing effort. "It does stick in one's craw a bit," Wilson remarks. But Gilman asserts that Celera's revised plan simply makes good business sense. "The point is not just to sit around and sequence for the rest of our lives," Gilman adds. "So, yes, we'll use our [threefold] coverage to order the public data, and that will give us what we believe to be a very accurate picture of the human genome." In early May the HGP announced it had completed its own working draft as well as a finished sequence for chromosome 21, which is involved in Down's syndrome and many other diseases. (For a full account of the chromosome 21 story, go to www.sciam.com/explorations/2000/051500chrom21 on the World Wide Web.)

Until now, the genome generators have focused on the similarities among us all. Scientists think that 99.9 percent of your genes perfectly match those of the person sitting beside you. But the remaining 0.1 percent of your genes vary—and it is these variations that most interest drug companies. Even a simple single-nucleotide polymorphism (SNP)—a T, say, in one of your gene sequences, where your neighbor has a C—can spell trouble. Because of these tiny genetic variations, Venter claims, many drugs work only on 30 to 50 percent of the human population. In extreme cases, a drug that saves one person may poison another. Venter points to the type II diabetes drug ezulin, which has been linked to more than 60 deaths from liver toxicity worldwide. "In the future, a simple genetic test may determine whether you're likely to be treated effectively by a given drug or whether you face the risk of being killed [by that same drug]" Venter predicts. While fleshing out its rough genome, Celera has also been comparing some of the genes with those from other individuals, building up a database of SNPs (pronounced "snips").

Other companies, too, hope to cash in on pharmacogenomics. Drug giants are partnering with smaller genomics-savvy companies to fulfill their gene dreams: Pfizer in New York City has paired with Incyte Genomics in Palo Alto, California; SmithKline Beecham in Philadelphia has ties to Human Genome Sciences in Rockville; and Eli Lilly in Indianapolis has links to Millennium Pharmaceuticals in Cambridge, Massachusetts. At this point, personalized medicine is still on the lab bench, but some business analysts say it could become an $800 million marker by 2005. As Venter puts it: "This is where we're headed."

But the road is sure to be bumpy. One sticking point is the use of patents. No one blinks when Volvo patents a car design or Microsoft patents a software program, according to John J. Doll, director of the U.S. Patent and Trademark Office's biotechnology division. But many people are offended that biotechnology companies are claiming rights to human DNA—the very stuff that makes us unique. Still, without such patents, a company like Myriad Genetics in Salt Lake City couldn't afford the time and money required to craft tests for mutations in the genes *BRCA1* and *BRCA2*, which have been linked to breast and ovarian cancer. "You simply must have gene patents," Doll states.

Most scientists agree, although some contend that companies are abusing the public genome data that have been so exactingly sequenced—much of them with federal dollars. Dutifully reporting their findings in GenBank, HGP scientists have offered the world an unparalleled glimpse at what makes a human. And Celera's scientists aren't the only ones peering in—in April, GenBank logged roughly 35,000 visitors a day. Some work at companies like Incyte, which mines the public data to help build its own burgeoning catalogue of genes—and patents the potential uses of those genes. Incyte has already won at least 500 patents on full-length genes—more than any other genomics company—and has applied for roughly another 7,000 more. Some researchers complain that such companies are patenting genes they barely understand and, by doing so, restricting future research on those genes. "If data are locked up in a private database and only a privileged few can access it by subscription, that will slow discovery in many diseases," warns Washington University's Wilson.

Incyte president Randal W. Scott, however, sees things differently: "The real purpose of the Human Genome Project is to speed up research discoveries, and our work is a natural culmination of that. Frankly, we're just progressing at a scale that's beyond what most people dreamed of." In March 2000 Incyte launched an e-commerce genomics program—like an Amazon.com for genes—that allows researchers to order sequence data or physical copies of more than 100,000 genes on-line. Subscribers to the company's genomics database include drug giants such as Pfizer, Bayer and Eli Lilly. Human Genome Sciences has won more than 100 gene patents—and filed applications for roughly another 7,000—while building its own whopping collection of genes to be tapped by its pharmaceutical partners, which include SmithKline Beecham and Schering-Plough.

The federal government has added confusion to the patent debate. Last March, President Bill Clinton and British prime minister Tony Blair released an ambiguous statement lauding open access to raw gene data—a comment some news analysts interpreted as a hit to Celera and other genomics companies that have guarded their genome sequences carefully. Celera and the HGP consortium have sparred over the release of data, chucking early talks of collaboration when the company refused to release its gene sequences immediately and fully into the public domain. The afternoon Clinton and Blair issued their announcement, biotech stocks slid, with some dropping 20 percent by day's end. A handful of genomics companies scrambled to set up press conferences or issue statements that they, indeed, did make available their raw genome data for free. In the following weeks, Clinton administration officials clarified that they still favor patents on "new gene-based health care products."

The sticky part for most patent seekers will be proving the utility of their DNA sequences. At the moment, many patent applications rely on computerized prediction techniques that are often referred to as "in silico biology." Armed with a full or partial gene sequence, scientists enter the data into a computer program that predicts the amino acid sequence of the resulting protein. By comparing this hypothetical protein with known proteins, the researchers take a guess at what the underlying gene sequence does and how it might be useful in developing a drug, say, or a diagnostic test. That may seem like a wild stab at biology, but it's often enough to win a gene patent.

"We accept that as showing substantial utility," Doll says. Even recent revisions to federal gene-patent standards—which have generally raised the bar a bit on claims of usefulness—ask only that researchers take a reasonable guess at what their newfound gene might do.

Testing, Testing

Patents have already led to more than 740 genetic tests that are on the market or being developed, according to the National Institutes of Health. These tests, however, show how far genetics has to go. Several years after the debut of tests for *BRCA1* and *BRCA2*, for instance, scientists are still trying to determine exactly to what degree those genes contribute to a woman's cancer risk. And even the most informative genetic tests leave plenty of questions, suggests Wendy R. Uhlmann, president of the National Society of Genetic Counselors. "In the case of Huntington's, we've got a terrific test," Uhlmann avers. "We know precisely how the gene changes. But we can't tell you the age when your symptoms will start, the severity of your disease, or how it will progress."

Social issues can get in the way, too. After Kelly Westfall's mother tested positive for the Huntington's gene, Westfall, age 30, immediately knew she would take the test as well. "I had made up my mind that if I had Huntington's, I didn't want to have kids," declares Westfall, who lives in Ann Arbor, Michigan. But one fear made her hesitate: genetic discrimination. Westfall felt confident enough to approach her boss, who reassured her that her job was safe. Still, she worried about her insurance. Finally, rather than inform her insurer about the test, Westfall paid for it—some $450, including counseling—out of pocket. (To her relief, she tested negative.)

The HGP's Collins is among those calling for legislation to protect people like Westfall. A patchwork of federal and state laws are already in place to ban genetic discrimination by insurers or employers, but privacy advocates are lobbying Congress to pass a more comprehensive law. In February 1999 President Clinton signed an executive order prohibiting all federal employers from hiring, promoting or firing employees on the basis of genetic information. It remains to be seen whether private companies will follow suit.

In the meantime, Celera is now ready to hawk its human genome, complete with crib notes on all the genes, to on-line subscribers worldwide. "It's not owning the data—it's what you do with it," Venter remarks. He envisions a Celera database akin to Bloomberg's financial database or Lexis-Nexis's news archives, only for the genetics set. Which 300 genes are associated with hypertension? What, exactly, does each gene do? These are the kinds of queries Celera's subscribers might pose—for a price. As of this writing, Celera planned to offer a free peek at the raw genome data on-line, but tapping into the company's online toolkit and full gene notes will cost corporate subscribers an estimated $5 million to $15 million a year, according to Gilman. Academic labs will pay a discounted rate: $2,000 to $15,000 a year.

Internet surfers can now visit GenBank for free. With all this information available, will scientists really pay Celera? Venter thinks so. "We just have to have better tools," he says. For genomics, that is becoming a familiar refrain.

The Other Genomes

Julia Karow

What do we have in common with flies, worms, yeast and mice? Not much, it seems at first sight. Yet corporate and academic researchers are using the genomes of these so-called model organisms to study a variety of human diseases, including cancer and diabetes.

The genes of model organisms are so attractive to drug hunters because in many cases the proteins they encode closely resemble those of humans—and model organisms are much easier to keep in the laboratory. "Somewhere between 50 and 80 percent of the time, a random human gene will have a sufficiently similar counterpart in nematode worms or fruit flies, such that you can study the function of that gene," explains Carl D. Johnson, vice president of research at Axys Pharmaceuticals in South San Francisco.

Here's a rundown on the status of the genome projects of the major model organisms today:

The Fruit Fly

The genome sequence for the fruit fly *Drosophila melanogaster* was completed this past March by a collaborative of academic investigators and scientists at Celera Genomics in Rockville, Maryland.

The researchers found that 60 percent of the 289 known human disease genes have equivalents in flies and that about 7,000 (50 percent) of all fly proteins show similarities to known mammalian proteins.

One of the fly genes with a human counterpart is p53, a so-called tumor suppressor gene that when mutated allows cells to become cancerous. The p53 gene is part of a molecular pathway that causes cells that have suffered irreparable genetic damage to commit suicide. In March 2000 a group of scientists, including those at Exelixis in South San Francisco, identified the fly version of p53 and found that—just as in human cells—fly cells in which the p53 protein is rendered inactive lose the ability to self-destruct after they sustain genetic damage and instead grow uncontrollably. Similarities such as this make flies "a good trade-off" for studying the molecular events that underlie human cancer, according to one of the leaders of the fly genome project, Gerald M. Rubin of the Howard Hughes Medical Institute at the University of California at Berkeley: "You can do very sophisticated genetic manipulations [in flies] that you cannot do in mice because they are too expensive and too big."

The Worm

When researchers deciphered the full genome sequence of the nematode *Caenorhabditis elegans* in 1998, they found that roughly one-third of the worm's proteins—more than 6,000—are similar to those of mammals. Now several companies are taking advantage of the tiny size of nematodes— roughly one millimeter—by using them in automated screening tests to search for new drugs.

To conduct the tests, scientists place between one and 10 of the microscopic worms into the pill-size wells of a plastic microtiter plate the size of a dollar bill. In a version of the test used to screen for diabetes drugs, the researchers use worms that have a mutation in the gene for the insulin receptor that causes them to arrest their growth. By adding various chemicals to the wells, the scientists can determine which ones restore the growth of the worms, an indication that the compounds are bypassing the faulty receptor. Because the cells of many diabetics no longer respond to insulin, such compounds might serve as the basis for new diabetes treatments.

The Yeast

The humble baker's yeast *Saccharomyces cerevisiae* was the first organism with a nucleus to have its genetic secrets read, in 1996. Approximately 2,300 (38 percent) of all yeast proteins are similar to all known mammalian proteins, which makes yeast a particularly good model organism for studying cancer: scientists first discovered the fundamental mechanisms cells use to control how and when they divide using the tiny fungus.

"We have come to understand a lot about cell division and DNA repair—processes that are important in cancer—from simple systems like yeast," explains Leland H. Hartwell, president and director of the Fred Hutchinson Cancer Research Center in Seattle and cofounder of the Seattle Project, a collaboration between academia and industry. So far Seattle Project scientists have used yeast to elucidate how some of the existing cancer drugs exert their function. One of their findings is that the common chemotherapeutic drug cisplatin is particularly effective in killing cancer cells that have a specific defect in their ability to repair their DNA.

The Mouse

As valuable as the other model organisms are, all new drugs must ultimately be tested in mammals—and that often means mice. Mice are very close to humans in terms of their genome: more than 90 percent of the mouse proteins identified so far show similarities to known human proteins. Ten laboratories across the U.S., called the Mouse Genome Sequencing Network, collectively received $21 million from the National Institutes of Health last year to lead an effort to sequence the mouse genome. They have completed approximately 3 percent of it, and their goal is to have a rough draft ready by 2003. But that time line might be sped up: Celera announced in April that it is turning its considerable sequencing power to the task.

Designer Genomes

Karen Hopkin

"Life! I've created LIFE!" shrieks the crazed scientist, eyes wild, hair spiking every which way, deep in the throes of megalomania. The scene is recognizable at once as the melodramatic centerpiece of many a late-night sci-fi flick, both good and bad.

What's more incredible is that such a scene may be playing itself out in a real lab sometime soon. The main difference—aside from the fact that most scientists now comb their hair—will be the creature on the table. Rather than a hulking monster made of body parts pilfered from a graveyard and stitched together by some scientist's fawning lackey, the artificial organism will be a bacterium—a microscopic life-form 1,000 times smaller than the smallest grain of sand.

Spurring this revolution is a new kind of recipe book: in the past five years researchers have determined the complete genomes—the exact sequences of the thousands of nucleotide base pairs that make up the DNA—of 24 different organisms, including yeast and the common intestinal bacterium *Escherichia coli*. As they examine and compare these simple genome sequences, investigators are gaining a fuller understanding of the fundamental instructions for life. Many believe the day is not far off when they will be able to design and create entirely new organisms—new life—from scratch.

Of course, scientists have been engaged in some form of genetic engineering—introducing single genes into the DNA of microorganisms such as *E. coli*—since the 1970s. They have tweaked bacteria into producing human proteins, engineered corn plants that can make pesticides and grown tobacco plants that clean up mercury from the soil. What makes *genome* engineering different is the scale: researchers are now beginning to outfit microorganisms with new biochemical pathways involving dozens of

"Our daughter cell may have my ability to take up inorganic ions, but she's got your wonderful talents at amino acid metabolism."

FUNCTION OF ESSENTIAL GENES	NUMBER OF GENES
Translation: Assembly of amino acids into a protein, based on the blueprint provided by the sequence of nucleotides in a molecule of messenger RNA	95
Energy: Production of enzymes necessary to allow the microbe to extract energy from nutrients such as simple sugars	34
Nucleotide Metabolism: Synthesis or recycling of the four chemical bases that make up a strand of DNA or RNA	23
Replication: Creation of a duplicate copy of the bacterial DNA chromosome, without which the microorganism could not reproduce	18
Chaperones: Production of molecules that guide, or "chaperone," the correct assembly of newly produced proteins	13
Transcription: Conversion of a strand of DNA into a sequence of RNA, from which a protein could be manufactured	9
Recombination and Repair: Detection and repair of breaks or errors that can occur in replicating DNA for reproduction	8
Coenzyme Metabolism: Synthesis and use of small-molecule co-factors that help some proteins to perform their tasks	8
Exopolysaccharides: Production of complex sugars that form part of the cell wall or external capsule	8
Amino Acid Metabolism: Synthesis or scavenging of the amino acids that are the building blocks of proteins	7
Lipid Metabolism: Production of lipids that store energy and form the bulk of the cell membrane	6
Uptake of Inorganic Ions: Production of the channels that permit the cell to respond to changes in its environment and to import salts and metals	5
Secretion AND receptors: Synthesis of molecules that enable cells to export proteins and respond to external signals such as the presence of nutrients	5
Other conserved proteins: Synthesis of additional proteins or RNAs with essential but as yet unknown functions	18

BASIC GENES: By comparing the genomes of the microbes Hemophilus influenzae (1,700 genes) and Mycoplasma genitalium (500 genes), scientists may have determined the 257 genes essential for life, at least for microbes.

genes packaged in long stretches of DNA, thereby altering extended segments of the microbes' genomes. Information obtained from the federally sponsored Human Genome Project and other genome-sequencing efforts provides genome engineers with the necessary raw materials—genes and the DNA sequences that control them—as well as a better blueprint of how organisms are put together.

Genome engineering will enable scientists to design microbes that can perform just about any biochemical task—synthesizing increasingly complex molecules or breaking them down. Imagine bugs custom-made to whisk away the "bioorganic halogenated compounds that cover half of New Jersey," says Roger Brent of the Molecular Sciences Institute in Berkeley, California.

Engineered microbes may even make molecular electronics a reality, suggests Gerald J. Sussman, a computer scientist and engineer at the Massachusetts Institute of Technology. When computer parts are reduced to the size of single molecules, industrious microbes could be directed to lay down complex electronic circuits. "Bacteria are like little workhorses for nanotechnology; they're wonderful at manipulating things in the chemical and ultramicroscopic worlds," Sussman says. "You could train them to become electricians and plumbers, hire them with sugar and harness them to build structures for you."

How will genome engineers build these marvelous microbial machines? Many will simply modify an existing creature by adding a biochemical pathway cobbled together from other organisms. But even that is a daunting task. Tailoring an existing system to suit one's needs requires quite a bit of knowledge about the pathway: Which steps are slowest? Where are the most likely bottlenecks? Genome engineers are turning to computer modeling to help design and test their system.

"We want to learn to program cells the same way as we program computers," says Adam P. Arkin, a physical chemist at Lawrence Berkeley National Laboratory. Some genome engineers have started by building the biological equivalent of the most basic switch in a computer—a digital flip-flop. Such a cellular toggle switch—made of DNA and some well-characterized regulatory proteins—might be devised to turn on a specific gene when exposed to a particular chemical.

These switches could be turned into sensitive biosensors for warning devices that would light up when they detect bioterrorist weapons such as botulin toxin or anthrax spores, according to James J. Collins, a physicist and bioengineer at Boston University. They could also be used in gene therapy: implanted genes might be controlled with single doses of specially selected drugs—one to switch the gene on, another to switch it off.

"It sounds simple," says Eric Eisenstadr of the Office of Naval Research (ONR), an agency that sponsors such projects. "But believe it or not, it isn't that easy to do." Selecting the appropriate genes—and configuring them to produce the desired response—is tricky business. Even so, Eisenstadt predicts that such genetic switches will be the "first baby steps" on the way to designing new regulatory pathways and eventually novel organisms.

Genome engineers trying to make such switches at least have a pattern to copy; nature serves as both teacher and supplier. "Cells switch genes on and off all the time," observes M.I.T.'s Thomas F. Knight, Jr., a computer scientist turned bioengineer. By taking advantage of nature's designs, genome engineers are starting off with circuits and components that have been "evolutionarily validated" as parts that work well, Brent adds.

Some researchers are harnessing the powers of evolution even more directly. They are using the principle of natural selection (in this case, survival of the fittest) to generate improved enzymes and perhaps whole organisms. In a process described as DNA shuffling, Willem P. C. Stemmer and his colleagues at Maxygen in Redwood City, California, isolate the genes for a particular enzyme from a handful of microbes. They break the genes into fragments and randomly introduce mutations to provide added variety. Then they shuffle and stitch the fragments back together.

By then screening for the mutant enzyme that is the fastest or most stable, investigators wind up with a hybrid that might be thousands of times more efficient than any of its parent enzymes, says Maxygen's Jeremy Minshull. Stemmer and his colleagues plan to apply a similar technique to shuffling not just single genes but whole genomes, which should yield bacteria optimized for whatever properties they desire—the ability to detoxify New Jersey, for example.

Andrew D. Ellington and his associates at the University of Texas use selective pressures to steer bacteria toward something even more unnatural—accepting and using amino acids that do not occur in nature and that are normally poisonous to living organisms. Ellington hopes that these funky bugs, which he calls Un-coli, will perform novel chemical reactions. Such as? "We don't know," he chirps with glee. "That's what makes this fun."

Rather than tinkering with existing bacteria, other scientists are talking seriously about building a creature from scratch, the ultimate engineering feat. Their approach is to start small, and several groups of investigators are trying to determine the minimal set of genes necessary for a cell to survive and reproduce.

One way to ascertain which genes are essential for life is to examine those present in microbes that have been fully sequenced and see which ones nature has elected to preserve. Eugene V. Koonin and Arcady R. Mushegian of the National Institutes of Health's National Center for Biotechnology Information have done just that. They compared two fully sequenced microbes: *Hemophilus influenzae*, with 1,700 genes, and *Mycoplasma genitalium*, with 500 genes—the smallest bacterial genome sequenced to date.

Koonin and Mushegian conclude that only 250 or so genes are required for life. J. Craig Venter and his colleagues at the Institute for Genomic Research (TIGR)—the team that sequenced *H. influenzae* and *M. genitalium*—venture that it's closer to 300. An organism with these 250 or 300 genes—whatever they are—would be able to perform the dozen or so functions required for life: manufacturing cellular components such as DNA, RNA, proteins and fatty acids; generating energy; repairing damage; transporting salts and other molecules; responding to chemical cues in their environment; and replicating. Although each of these functions requires multiple genes, the whole shebang could be carried in a genome some 300,000 nucleotide bases in length—about half the size of *M. genitalium*'s.

To determine which genes are truly indispensable, some researchers have been deleting them one by one. Venter's TIGR team is knocking genes out of *M. genitalium*. Other groups are performing similar elimination experiments in *E. coli* and yeast. Pharmaceutical companies are using *E. coli* mutants generated by George M. Church of Harvard Medical School to identify new targets for antibiotics—genes that appear to be essential for bacteria but are not found in humans.

Knowing which genes are necessary is one thing, but how do you turn that information into life? Today's DNA synthesizers are not capable of whipping up genome-size chunks of DNA. But researchers are working on techniques for synthesizing large rings of DNA that hold the genes for a single biochemical function—say, all the enzymes necessary to produce ATP, the molecule that cells use for energy. Glen A. Evans and his colleagues at the University of Texas Southwestern Medical Center can churn out DNA 10,000 to 20,000 nucleotide bases in length; they'd like to make pieces 10 times as long. With the proper genes in hand, all that would remain would be for scientists to stuff the pieces of DNA into an empty cell sac—most likely an animal cell from which the nucleus had been removed. The proteins left in the gutted cell, Evans and others hope, would begin making the molecules necessary to jump-start this new form of life.

Of course, producing novel life-forms will raise many concerns, from ecological to ethical. Potential problems have already surfaced in the genetically engineered plants of today. For example, corn that produces its own insecticide may kill harmless bugs (like monarch butterflies). Minimal-genome microbes, however, might not even be able to survive outside the lab. "I doubt this minimal life-form will be lurching around frightening villagers," comments Thomas H. Murray of the Hastings Center for Bioethics.

Then there is the philosophical question: If scientists can actually create life, are they playing God? "People usually raise that point as a way to forestall discussion of the real issues," says David Magnus of the University of Pennsylvania Center for Bioethics. He and his colleagues have been considering the ethical implications of synthesizing cells from the ground up—an event he guesses will grab headlines in the next five years. After two years of contemplation, the group has concluded that the potential benefits of engineering life—which Magnus says include better gene therapy techniques and an enhanced understanding of cell biology—outweigh the possible dangers. But these issues, he asserts, should be addressed by scientists and society.

That discussion had better start soon, because genome engineers are closer than even most scientists realize to making creatures unlike anything ever seen on Earth. What this brave new bioengineered world will look like is hard to say. "But it's going to be awesome," ONR's Eisenstadt predicts. "I mean, it's life."

Suggested Reading

Green, E. "The Human Genome Project and Its Impact on the Study of Human Disease." In *Metabolic and Molecular Bases of Inherited Disease.* Charles R. Scriver, 8th ed. Boston: McGraw-Hill, 2000.

Pennisi, E. "Are Sequencers Ready to Annotate the Human Genome?" *Science* 287, no. 5461 (March 24, 2000): 2183.

Chapter 5

What Can We Do with Gene Therapy?

Gene Therapy

Just a year ago, genetic therapies—treatments that work by rewriting bits of genetic code in a patient's cells—were widely heralded as the next great champion of modern medicine. Then the champ hit an unexpected slump. Gene therapy took a standing eight-count last winter, after drug contenders sponsored by a host of biotechnology and drug companies failed to cure a single patient of disease. In a highly critical report issued in December 1995, a review panel at the National Institutes of Health chided researchers and investors for rushing treatments into human clinical trials before fully understanding all the natural defenses that genetic medicines must conquer or evade if they are to work.

Somewhat chastened, geneticists are studying their failures and starting to develop a clearer picture of what they are up against. Many researchers are optimistic that the present retrenchment actually bodes well for the long-term success of genetic medicines. In laboratories and on Wall Street, there are signs that gene therapy is starting to stage a comeback.

The challenge facing genetic medicines is daunting. First, they must somehow deliver their genetic payload into enough cells to do some good. Retroviruses seemed well suited for this task, because these kinds of viruses normally infect cells by copying part of their DNA into the genetic code of a host cell. Most early trials of genetic medicines therefore co-opted retroviruses, replacing their harmful parts with genes intended to help treat a disease, such as cystic fibrosis or brain cancer.

But viral drugs can take effect only if they can slip past the multilayered defenses of the human immune system. First comes an onslaught of antibodies soon after any familiar virus is detected in the bloodstream. These antibodies quickly bind up the virus and can also cause side effects such as inflammation. Viral particles that make it to target cells face a second obstacle: a tough membrane shielding the cell's DNA from attackers. Finally, those retroviruses that are lucky enough to make it past the immune defenses and to infect cells do so in an unpredictable manner; they typically will insert the therapeutic gene at a random position in the cell's DNA. The new gene might interrupt an important sequence, actually harming the cell. Even in the best case, new genes often end up in dormant stretches of DNA, where they do not get switched on frequently enough to make much of a difference to the patient.

Geneticists were humbled by these barriers, but they were not stumped. A second wave of enthusiasm for gene therapy is now well under way, thanks to recent advances that suggest new strategies. In September 1996, RPR Gencell (a network of gene therapy research centers organized by the French company Rhone-Poulenc Rorer) published results in *Nature Medicine* describing its test of a retroviral gene therapy for lung cancer. The researchers injected the drug containing normal versions of p53—a gene that suppresses tumors—directly into nine patients' tumors. This technique avoided triggering a general immune response and exploited the rapid division of tumor cells. Tumors shrank significantly in three of the patients and stopped growing in three others; nevertheless, all nine patients died.

Results from two other groups recently suggested that it might be possible to design gene therapies that altogether avoid viruses and their many drawbacks. Workers at the University of Chicago and at Vical, a biotechnology firm in San Diego, rolled a gene for erythropoietin into a circular DNA package called a plasmid. Erythropoietin is a hormone that triggers the body to produce red blood cells. Another biotechnology company, Amgen, sells nearly $1 billion of its synthetic version each year to patients afflicted with anemia and other blood disorders.

In an article published October 1, 1996 in the *Proceedings of the National Academy of Sciences*, the Chicago group reported that simply injecting the naked DNA into the hindquarters of normal mice boosted their red blood cell counts by a third. Equally important, the counts remained higher up to 90 days after the injection, strongly indicating that the genes had taken hold (at least for a while) and begun producing hormone. "Our results suggest that intramuscular injection of currently available genes could be used to treat a variety of serum protein deficiency diseases," such as anemia, hemophilia and diabetes, says Jeffrey M. Leiden, director of the study at the University of Chicago.

Boosting blood-cell production does little good for patients whose blood cells are malformed, such as those of sickle-cell anemics. The ultimate goal of gene therapy is not to compensate for genetic diseases but to erase them completely. Preliminary work published in the September 6, 1997, issue of *Science* offers a reason to hope that goal may be possible. A team led by Allyson Cole-Strauss and Kyonggeun Yoon of Thomas Jefferson University in Philadelphia experimented on cells containing a mutant gene that causes sickle-cell anemia. To make their genetic drug, they combined DNA for the normal version of this gene with RNA for the same gene.

When they injected the drug into the diseased cells, the RNA/DNA particles homed in on the particular stretch of the genome that matched their codes and formed triple-stranded DNA that covered the mutation. The cells' normal DNA-repair machinery then apparently replaced the mutation with the normal code—thus permanently curing 10 to 20 percent of the cells. The researchers still have to demonstrate that this technique works in human cells and in human bodies.

While scientists and the public slowly allow optimism to creep back into discussions about gene therapy, investors and biotechnology companies are unabashedly bullish once more. By July, 216 clinical trials of gene therapies were planned or under way, according to the Pasteur Institute in Paris. "Biotech firms of every kind are scrambling to reposition themselves as genomics companies," says Joan E. Kureczka, an industry publicist.

Gene therapy may emerge a winner in this round, though the match will likely draw on for years to come. The longer the search and the larger the investment, however, the more expensive treatments will be when they do arrive. The greatest challenges to the medical wonders promised by gene therapy may well turn out to be economic rather than scientific.

—*W. Wayt Gibbs, editor,* Scientific American

Overcoming the Obstacles to Gene Therapy

Theodore Friedmann

In the late 19th century, when the pioneering architect Daniel H. Burnham was planning some of the first modern skyscrapers, his associates were skeptical about erecting buildings that soared into the clouds. Burnham reportedly warned the skeptics against making "little plans," having "no magic to stir men's blood." He urged them to reach beyond traditional architectural boundaries, to think once inconceivable thoughts and to perform previously unimagined deeds—the hallmarks of revolutions.

Revolutionary changes have also occurred in medicine over the past few centuries. Witness the new understandings and practices that issued from the introduction of microscopy, anesthesia, vaccination, antibiotics and transplantation. Medicine is now preparing to undergo another epochal shift: to an era in which genes will be delivered routinely to cure or alleviate an array of inherited and acquired diseases.

Preparing for a radical change, yes, but not yet in the midst of it. By emphasizing hopes and downplaying uncertainties, some overzealous researchers, representatives of industry and members of the lay and scientific media have implied that gene therapy is already advanced enough for widespread application. It is not.

Arguably, the conceptual part of the gene therapy revolution has indeed occurred. Whenever a new gene is discovered, researchers and nonscientists immediately ask whether it can be used to treat some disorder, even when more traditional approaches might be applied. But the technical part of the revolution—the ability to correct disease—is another story. Investigators have accomplished the requisite first steps: they have shown that transferred genes can be induced to function in the human body, at times for several years. So far, however, no approach has definitively improved the health of a single one of the more than 2,000 patients who have enrolled in gene therapy trials worldwide.

This lack of a convincing therapeutic benefit is sobering. Yet it would be a mistake to doubt gene therapy's powerful future. Remember, the field is young; in the U.S., trials in patients have been carried out for fewer than 10 years. A more realistic interpretation of the unspectacular clinical results thus far is that they reflect researchers' imperfect initial gropings toward a difficult new technology and that the obstacles are more formidable than many of us had expected.

A central challenge, as a federally commissioned critique of the gene-therapy research effort noted in 1995, is perfecting methods for delivering therapeutic genes to cells. Often genes introduced into patients do not reach enough of the appropriate cells or, for reasons that are not always clear, function poorly or shut off after a time. Under those conditions, a gene that could potentially be helpful would have little chance of affecting a disease process.

In this article I will outline some of the most pressing technological stumbling blocks to successful gene transfer and the strategies being considered to cope with those difficulties. I will deal only with therapy affecting somatic cells, the kinds that are neither sperm nor egg. To date, research aimed at human gene therapy has avoided manipulations that would deliberately affect descendants of the treated individuals, perhaps in unintended ways. The need for enlightened public debate over the merits and risks of germ-line therapy has, however, been made more urgent by the cloning of an adult sheep.

How Genes and Gene Therapy Work

Anyone who wants to understand the obstacles to gene therapy should first know a bit about what genes do and about how attempts at gene therapy are currently carried out. An individual gene in the human cell is a stretch of DNA that, in most cases, acts as a blueprint for making a specific protein; it spells out the sequence of amino acids composing that protein. All cells in a body carry the same genes in the chromosomes of the nucleus. But neurons, say, behave unlike liver cells because different cells use, or express, distinct

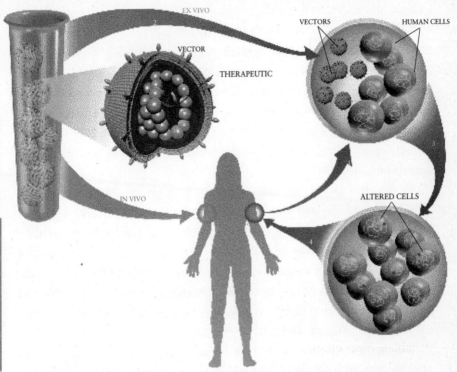

Delivery of genes to human subjects is sometimes accomplished directly by putting vectors (agents carrying potentially therapeutic genes) straight into some target tissue in the body (in vivo). More often the ex vivo approach is used: physicians remove cells from a patient, add a desired gene in the laboratory and return the genetically corrected cells to the patient. An in vivo approach still in development would rely on "smart" vectors that could be injected into the bloodstream or elsewhere and would home to specific cell types anywhere in the body.

subsets of genes and hence make separate sets of proteins (the main functionaries of cells). Put more precisely, each cell copies only selected genes into individual molecules of messenger RNA, which then serve as the templates from which proteins are constructed.

If a particular gene is mutated, its protein product may not be made at all or may work poorly or even too aggressively. In any case, the flaw may disturb vital functions of cells and tissues that use the normal gene product and can thereby cause symptoms of disease.

Overcoming the Obstacles to Gene Therapy

Historically, physicians have treated disorders stemming from inherited genetic mutations not by altering genes but by intervening in the biological events resulting from a mutation. For example, dietary restriction has long been prescribed for phenylketonuria, in which loss of a gene leads to the toxic buildup of the metabolic products of the amino acid phenylalanine. Unfortunately, nongenetic manipulations are usually only partly effective against inherited ills.

In the early 1970s this fact—combined with growing understanding of how genes function and with the discovery of the genes underlying many inherited ills—led to the suggestion that better results might be achieved by attacking inborn diseases at their source. Among the genetic diseases that have been studied are cystic fibrosis (which mainly affects the lungs), muscular dystrophy, adenosine deaminase deficiency (which severely impairs immunity), and familial hypercholesterolemia (which leads to the early onset of severe atherosclerosis).

Surprisingly, as time went on, it became clear that even acquired maladies often have a genetic component that can theoretically be a target of a genetic correction strategy. Indeed, quite unexpectedly, more than half of all clinical trials for gene therapy these days aim at cancer, which in most cases is not inherited but results from genetic damage accumulated after birth. A number of trials also focus on AIDS, which is caused by the human immunodeficiency virus (HIV).

In principle, a normal gene can be delivered so that it physically takes the place of a flawed version on a chromosome. In practice, such targeted insertion of a gene into a chromosome is not yet achievable in people; fortunately, it often is not required. Most attempts are gene therapy simply add a useful gene into a selected cell type or compensate for a missing or ineffective version or to instill some entirely new property. Many proposed anticancer gene therapies under study take this last tack: they aim to induce cancer cells to make substances that will kill those cells directly, elicit a potent attack by the immune system or eliminate the blood supply that tumors require for growth.

Some gene therapy groups are also devising strategies to compensate for genetic mutations that result in destructive proteins. In one approach, called antisense therapy, short stretches of synthetic DNA act on messenger RNA transcripts of mutant genes, preventing the transcripts from being translated into abnormal proteins. Related tactics deploy small RNA molecules called ribozymes to degrade messenger RNA copied from aberrant genes. A rather different plan provides a gene for a protein, called an intracellular antibody, that can block the activity of the mutant protein itself. Some therapeutic strategies rely on the design of hybrids of DNA and RNA that might direct the repair of mutant genes.

Genes are currently provided to patients in two basic ways. In both cases, the genes are usually first put into transporters, or vectors, able to deposit foreign genes into cells. In the more common method, scientists remove cells from a selected tissue in a patient, expose them to gene-transfer vectors in the laboratory (ex vivo) and then return the genetically corrected cells to the individual. Other times researchers introduce the vectors directly into the body (in vivo), generally into the tissue to be treated. Our ultimate goal, of course, is to deliver vectors into the bloodstream or other sites and to have them act like homing pigeons, finding their own way to the desired cells—say, to organs that are hard to reach or to hidden cancer deposits. No such targeted carriers are yet ready for testing in patients, but work toward that end is advancing quickly.

In the body, certain genes will be helpful only if their expression is regulated tightly: in other words, they must give rise to just the right amount of protein at the right times. Biologists have yet to achieve such precise control over foreign genes put into the body. For many gene therapy applications, however, exquisite regulation will not be essential. Nor will it always be necessary to put genes into the cells that are in need of fixing. Sometimes more accessible cell types (say, muscle or skin) might be turned into protein factories; these factories would release proteins needed by nearby cells or might secrete proteins into the bloodstream for transport to distant sites.

Retrovirus Vectors: Flaws and Fixes

The key to success for any gene therapy strategy is having a vector able to serve as a safe and efficient gene delivery vehicle. From the start, viruses—

which are little more than self-replicating genes wrapped in protein coats—have drawn the most attention as potential vectors. They are attractive because evolution has designed them specifically to enter cells and express their genes there. Further, scientists can substitute one or more potentially therapeutic genes for genes involved in viral replication and virulence. In theory, then, an altered, tamed virus should transfer helpful genes to cells but should not multiply or produce disease.

The viruses that have been examined most extensively are retroviruses, which splice copies of their genes permanently into the chromosomes of the cells they invade. Such an integrated gene is copied and passed to all future generations of those cells. In contrast, many other kinds of viruses do not integrate their genetic material into a host's chromosomes. Their genes generally function in the body more transiently—in part because the genes do not replicate when recipient cells divide.

One group of ideal target cells for retrovirus vectors consists of so-called stem cells, which persist indefinitely and also produce more specialized descendant cells. Blood-forming stem cells, for example, give rise to every other type of blood cell (red cells, white cells of the immune system, and so on) and reconstitute the blood as needed; they also make more copies of themselves. At the moment, however, it is extremely difficult to identify human stem cells and modify them in safe, predictable ways.

Despite the appeal of retroviruses, which were first introduced as vectors in the early 1980s, they pose several challenges. They are promiscuous, depositing their genes into the chromosomes of a variety of cell types. Such lack of fine specificity for host cells can militate against direct delivery of the vectors into the body; uptake by cells that were not intended to receive the foreign gene could reduce transfer to the targeted population and might have unwanted physiological effects. Conversely, the retroviruses now receiving the most study fail to transfer genes to cell types that cannot divide or that do so only rarely (such as mature neurons and skeletal muscle cells). Current retrovirus vectors reach chromosomes only when the membrane surrounding the nucleus of the host cell dissolves, an event that occurs solely during cell division.

Also problematic is the fact that retroviruses splice their DNA into host

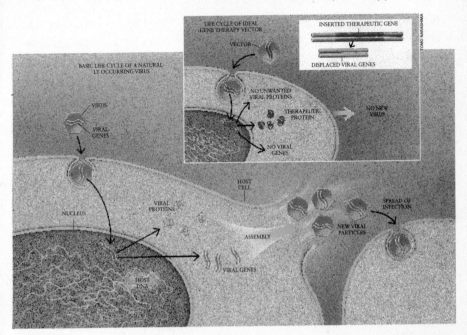

A naturally occurring virus (main panel) releases its genetic material into cells. Whether or not the genes become integrated into the DNA of the infected cell, they soon direct the synthesis of new viral particles that can injure the cell and infect others. To convert a wild-type virus into a safe gene therapy vector, scientists replace viral genes with ones specifying therapeutic proteins (inset), while ideally leaving only the viral elements needed for gene expression. Such vectors should enter cells and give rise to helpful proteins but should not multiply.

chromosomes randomly, instead of into predictable sites. Depending on where inserted genes land, they might disrupt an essential gene or alter genes in ways that favor cancer development. Tumors would probably result only rarely, but even the remote chance of increasing cancer risk must be taken seriously.

Researchers have made good progress recently in confronting the shortcomings of retroviruses as gene delivery vehicles. For instance, to increase specificity and thus enable retrovirus vectors to direct themselves to particular cells in the body, researchers are altering the viral envelope (the outermost surface). Like other viruses, retroviruses deposit their genetic cargo into a cell only if proteins projecting from their surface find specific mates, or receptors, on the cell. Binding of the viral proteins to the cellular receptors enables a retrovirus to fuse its envelope with the cell membrane and to release viral genes and proteins into the cell's interior. To make retroviruses more selective about the cells they invade, investigators are learning how to replace or modify natural envelope proteins or to add new proteins or parts of proteins to existing envelopes.

In an experiment showing that the replacement strategy is feasible, Jiing-Kuan Yee of the University of California at San Diego, with my laboratory at that university, substituted the envelope protein of the mouse leukemia virus with that of the human vesicular stomatitis virus. (The mouse virus, which causes no known disease in people, is the retrovirus that has been evaluated most extensively as a gene therapy vector.) The altered mouse retrovirus then infected cells bearing receptors for the human vesicular stomatitis virus instead of cells with receptors for the mouse virus.

Work on modifying existing envelope proteins is also proceeding well. Yuet Wai Kan and his colleagues at the University of California at San Francisco have recently linked a protein hormone to the envelope protein of the mouse leukemia virus. This hormone enabled the virus to infect human cells that displayed the receptor for that hormone.

Prospects for generating retrovirus vectors able to insert therapeutic genes into the chromosomes of nondividing cells are looking up as well. Inder M. Verma, Didier Trono and their colleagues at the Salk Institute for Biological Studies in San Diego have capitalized on the ability of HIV, a retrovirus, to deposit its genes into the nucleus of nondividing brain cells without waiting for the nuclear wrapping to dissolve during cell division.

The team removed genes that would allow HIV to reproduce and substituted a gene coding for a protein that was easy to trace. This vector then brought the traceable gene into nonreplicating cells, not only when the vector was mixed with cells in culture but also when it was injected directly into the brains of rats. HIV itself might one day prove to be a useful vector if worry that the disabled vectors might somehow become pathogenic can be allayed. Another tactic would transfer certain of HIV's useful genes—

particularly those coding for the proteins that transport genes to the nucleus—into retroviruses that do not cause human disease.

Finally, efforts are under way to ensure that retrovirus vectors will place genes less randomly into human chromosomes. Workers toiling in this taxing realm have recently been assisted by new understanding of how genes integrate into predictable sites in the DNA of other organisms, such as yeast.

Pros and Cons of Other Virus Vectors

Vectors derived from viruses other than retroviruses present their own sets of advantages and disadvantages. Those based on the ubiquitous human adenoviruses have gained the most popularity as alternatives to retroviruses in part because they are quite safe; the naturally occurring forms typically cause nothing more serious than chest colds in otherwise healthy people. Moreover, they infect human cells readily and, initially at least, tend to yield high levels of the therapeutic protein.

Adenovirus vectors dispatch genes to the nucleus but apparently do not insert them into chromosomes. This feature avoids the possibility of disturbing vital cellular genes or abetting cancer formation, but, regrettably for some applications, the genes are often effective only temporarily. Because the DNA eventually disappears, treatments for chronic conditions, such as cystic fibrosis, would have to be repeated periodically (every so many months or years). In some situations, though—say, when a protein is needed only temporarily to induce an immune response to cancer or to a pathogen—short-term expression of a foreign gene may be preferable. Another drawback, shared with retroviruses, is lack of specificity for target cells. As is true for retroviruses, however, scientists are rapidly devising ways to target adenovirus vectors to tissues of the researchers' choosing.

At the moment the more serious stumbling block to use of adenovirus vectors in patients is the body's strong immune response against them. During an initial round of treatment, such vectors might infect the appropriate cells and generate high amounts of the desired proteins. But soon host defenses come into play, killing the altered cells and inactivating their new genes. Further, once the immune system is alerted to the viruses, it eliminates them quickly if they are delivered a second time. Such responses probably have contributed to a shutdown of gene expression in a number of adenovirus gene-transfer studies in patients. Advancing understanding of the shortcomings of adenoviruses is now leading to a new generation of vectors that should reduce defensive interference. These enhancements have been achieved in part by removing or mutating the adenovirus genes most responsible for eliciting immune attacks.

Several other viruses are being explored as vectors as well—among them, adeno-associated viruses, herpesviruses, alphaviruses and poxviruses. None is perfected yet, but each is likely to have its own therapeutic niche. For example, adeno-associated viruses appeal because they cause no known diseases in people. What is more, in their natural form, they integrate their genes into human chromosomes. They are likely to be useful for some applications that now depend on retroviruses, but they are smaller and so may not be able to accommodate large genes. Herpesviruses, in contrast, do not integrate their genes into the DNA of their hosts. But they are attracted to neurons, some of which retain the viruses in a more or less innocuous state for the lifetime of the affected person. Herpesviruses therefore have potential as vectors for therapy aimed at neurological disorders.

Perfecting Nonviral Delivery Systems

As a group, vectors produced from viruses continue to show great promise, although researchers must always work to ensure that the viruses will not change in ways that will enable them to cause disease. This consideration and others have encouraged development of various nonviral methods for therapeutic gene transfer. In common with viruses, these synthetic agents generally consist of DNA combined with molecules that can condense the DNA, deliver it to cells and protect it from degradation inside cells. And, like virus vectors, they will almost certainly be used in medical practice eventually but are still in need of refinement. The genes transferred by nonviral vectors become integrated into the chromosomes of recipient cells in the laboratory but have done so only rarely after delivery into the body.

Whether lack of integration will be an advantage or disadvantage depends, as I have noted, on the particular goal of therapy.

Liposomes, which are small fatty (lipid) spheres, have been studied almost as long as retrovirus vectors. These synthetic bubbles can be designed to harbor a plasmid—a stable loop of DNA derived from bacterial viruses known as phages—in which original genes have been replaced by those intended to be therapeutic. Gene transfer by liposomes (or "lipoplexes," as current versions are increasingly called) is much less efficient than virus-mediated transfer but has advanced enough for these vectors to enter critical trials for such diseases as cancer and cystic fibrosis. Meanwhile alterations in the chemical composition of liposomes are addressing the efficiency problem and are beginning to produce vectors that mimic viruses in their targetability and prowess at gene transfer.

Newer kinds of vectors sheathe DNA in nonlipid coats. These coats include amino acid polymers and other substances intended to target therapeutic genes to the proper cells in the body and to protect the genes from being broken down by cellular enzymes. These complexes—studied intensively by Max Birnsteil and Matt Cotton of the institute of Molecular Pathology in Vienna and by David T. Curiel of the University of Alabama at Birmingham—have performed well in cell culture experiments. They are now being further modifiied and are undergoing testing in animal studies and in patients.

Some scientists are also exploring injecting so-called naked DNA—without a lipid wrapping—into patients. Initial results suggest that the naked-DNA strategy has exciting potential for immunization against infectious diseases, and even against certain kinds of cancer.

Alternatives to plasmids are being pursued as well. Notably, workers are learning to construct miniature chromosomes, or artificial human chromosomes, into which therapeutic genes can be spliced. These constructs will contain the minimum amount of genetic material needed to avoid degradation in the nucleus and loss during cell division. They will also incorporate elements that enable the artificial chromosomes to copy themselves accurately (and only once) each time a cell divides, just as ordinary chromosomes do.

Looking Ahead

In the future, as now, investigators will choose one or another gene delivery method on the basis of their therapeutic goal. If a patient inhereited a genetic defect and needs a continuing supply of the normal gene product throughout life, a vector that can integrate the therapeutic gene into the patient's chromosomes, where it will stay in perpetuity, might be best. Then a retrovirus or adeno-associate virus may be selected. If only short-term activity of a gene is needed, such as to arouse the immune system against cancer cells or an infectious agent, nonintegrating delivery vehicles, such as adenovirus vectors, liposomes or even naked DNA may be more suitable.

But the tools that finally come into common use almost certainly will not be the prototypes being tested today. And because no single technique will be perfect for every disorder, there will be many choices. The ideal gene transfer systems of the future will combine the best features of different vectors. Each system will be tailored to the specific tissue or cell type requiring modification, to the needed duration of gene action and to the desired physiological effect of the gene product. Scientists will also want to develop ways to alter the level of gene expression at will and to shut off or completely remove introduced genes if therapy goes awry.

Even when these gene delivery vectors are perfected, the challenges will not end. For instance, cells often modify foreign genes in ways that ultimately cause the genes to stop working. This activity is being addressed vigorously but is not yet solved. In addition, we still have few clues as to how defensive systems of patients will respond when they encounter a seemingly foreign protein from a therapeutic gene. To prevent an inactivating immune reactions, physicians might have to treat some patients with antirejection drugs or try to induce immune tolerance to the encoded protein by carrying out gene therapy very early in a patient's life (before the immune system is fully competent).

Although I have dwelled on certain technical challenges to gene therapy, I am nontheless highly optimistic that it will soon begin to prove helpful for some disease. Our tools are improving rapidly, and some of the burgeoning clinical trials clearly are on the verge of demonstrating real merit in ameliorating disease, even with today's imperfect techniques. Notably, it seems likely that gene-based immunotherapies for some malignancies, such as neuroblastoma and melanoma, will be shown convincingly in the next few years to slow the development of further disease and to force existing tumors to regress; they should then become helpful additions to existing therapies. But I must emphasize that it is only through insistence on rigorous science, carefully designed clinical studies and less exaggerated reporting of results that researchers can ensure timely, ethical and effective flowering of this exciting new field of medicine.

Gene Therapy for Cancer

R. Michael Blaese

In 1997 an estimated 1.38 million Americans will be newly diagnosed with cancer. Sadly, the main treatments currently available—surgery, radiation therapy and chemotherapy—cannot cure about half of them. This sobering fact has spurred serious efforts to develop additional strategies for treating the disease—ones based on the biology behind it. To that end, scientists are turning toward gene therapies, which involve introducing into the body genes that can potentially combat tumors. Researchers initially explored gene therapies for remedying conditions caused by defective genetic instructions, or mutations, passed on from one generation to the next. Most cancers are not inherited in this way but instead result from acquired mutations, produced by external factors such as tobacco smoke or high doses of radiation—or just pure bad luck. These mutations accumulate in cells over time, ultimately rendering the cells unable to control their own growth—an inability that leads to cancer [see "What You Need to Know about Cancer," special issue of *Scientific American*; September 1996].

Gene therapies in general deliver instructions—in the form of DNA sequences—to diseased cells so that they will produce a therapeutic protein of some kind. This type of therapy is possible because viruses, bacteria, plants and people all share the same genetic code. Researchers have learned a great deal in little time about how certain genes govern the fundamental processes of life and how they contribute to disease. And because genes from one species can be read and understood by another, experimenters can transfer genes between cells and species in their efforts to devise treatments.

For treating cancer, experimental gene therapies take varied forms: some involve imparting cancer cells with genes that give rise to toxic molecules. When these genes are expressed (that is, used by cells to make proteins), the resulting proteins then kill the cancer cells. Other designs aim to correct or compensate for acquired genetic mutations. Still others attempt to activate the processes by which such defects are normally repaired. And a host of ideas are coming from insights into how tumors evade recognition and destruction by the immune system, how they spread away from their sites of origin, how they gain a new blood supply and how they accomplish other feats that allow them to endure and spread.

Most of these approaches have yet to pass even the most preliminary clinical tests demonstrating their overall safety and efficacy, but these ideas may lead to better cancer treatments in the future.

Aside from promising actual treatments, gene therapy techniques have thus far helped physicians evaluate existing interventions. For example, in recent years doctors have increasingly relied on bone marrow transplants for treating cancers that fail to respond to traditional therapies. Frequently, the procedure is used for battling advanced stages of leukemia, a cancer that affects white blood cells, which are made by bone marrow. Before performing transplantation, oncologists take bone marrow from leukemia patients in remission. This apparently healthy bone marrow is stored away, and the patients are given superhigh doses of chemotherapy or radiation to kill off any residual cancer. Because these high doses destroy normal bone marrow, such an aggressive treatment would ordinarily kill the patient as well as the cancer. In a transplantation, though, the patient is "rescued" by receiving a transfusion of his or her own saved bone marrow.

This method should, in theory, cure leukemia. But sometimes the disease recurs anyway. Clinicians have often wondered what goes wrong. Do the high-dose radiation treatments sometimes fail to kill off all residual cancer cells, or are there sometimes undetected leukemia cells in the presumed disease-free marrow stored away during remission? To distinguish between these possibilities, scientists needed to devise a nontoxic and permanent tag so that they could mark the extracted bone marrow cells and find them later in the body. Malcolm K. Brenner of St. Jude Children's Research Hospital in Memphis did just that starting in late 1991, by inserting a unique sequence of bacterial DNA not found in humans into the patient's saved bone marrow. Brenner knew that if he detected this bacterial DNA in the recovering patient's blood and bone marrow after transplantation, it would prove that the "rescue marrow" was restoring the blood system. In addition, detection of this tag in recurrent cancer cells would prove that the rescue marrow was a source of leukemic cells.

This is in fact what Brenner and others have found in some cases, forcing a critical reappraisal of the use of bone marrow transplantation. For instance, it is now recognized that for certain types of cancer, it may be necessary to give additional treatments to the marrow itself to rid it of any contaminating cancer cells before transplantation. To that end, marrow "gene-marking" studies, identical to those described above, are helping to find the best solution. These studies allow researchers to compare sundry methods for purging the marrow of residual cancer. Different marker genes are used to tag marrow samples purged in different ways. As a result, physicians can determine how well marrow that has been purged in some way helps the patient following therapy. And, if the cancer recurs, they can also determine whether it has come from bone marrow that was purged in a particular way.

Gene Vaccinations

In terms of treatment, scientists have for more than three decades tried to find ways to sic the immune system on cancer—a tactic termed immunotherapy or vaccine therapy. And with good reason. Because immunity is a systemic reaction, it could potentially eliminate all cancer cells in a patient's body—even when they migrate away from the original tumor site or reappear after years of clinical remission. The problem with this strategy has been that the immune system does not always recognize cancer cells and single them out for attack. Indeed, many tumors manage to hide themselves from immune detection.

Recently, however, research in basic immunology has revealed means for unmasking such cancers. In particular, it now seems possible to tag cancer cells with certain genes that make them more visible to the immune system. And once awakened, the immune system can frequently detect even those cancer cells that have not been tagged.

The immune response involves many different cells and chemicals that work together to destroy in several ways invading microbes or damaged cells. In general, abnormal cells sport surface proteins, called antigens, that differ from those found on healthy cells. When the immune system is activated, cells called *B* lymphocytes produce molecules known as antibodies. These compounds patrol the body and bind to foreign antigens, thereby marking the antigen bearers for destruction by other components of the immune system. Other cells, called *T* lymphocytes, recognize foreign antigens as well; they destroy cells displaying specific antigens or rouse other killer *T* cells to do so. *B* and *T* cells communicate with one another by way of proteins they secrete, called cytokines. Other important accessory cells—antigen-presenting cells and dendritic cells—further help *T* and *B* lymphocytes detect and respond to antigens on cancerous or infected cells.

One gene therapy strategy being widely tested at the moment involves modifying a patient's cancer cells with genes encoding cytokines. First the patient's tumor cells are removed. Into these tumor cells, scientists insert genes for making cytokines, such as the *T* cell growth factor interleukin-2 (IL-2) or the dendritic cell activator called granulocyte-macrophage colony-stimulating factor (GM-CSF). Next, these altered tumor cells are returned to the patients's skin or muscle, where they secrete cytokines and thereby catch the immune system's attention. In theory, the altered cells should solicit vigorous immune cell activity at the site of the reinjected tumor. Moreover, the activated cells, now alerted to the cancer, could circulate through the body and attack other tumors.

In certain instances, these gene-modified tumor vaccines do seem to awaken the immune system to the presence of the cancer, and some striking clinical responses have been observed. All these clinical studies, however, are preliminary. In most cases, patient responses to these treatments have not been carefully compared with responses to conventional treatments alone. Also, the response patterns are not predictable, and they are not consistent from one tumor type to another or among patients who have the same type of cancer.

Another problem with these studies is that nearly every person tested so far has had widely disseminated terminal cancer. Usually these patients have previously received intensive anticancer therapy, which has weakened their immune systems. Thus, even if gene vaccines did activate immunity in these individuals, the responses might not be easily noticeable. Gene-modified tumor vaccines are most likely to prove beneficial in patients with minimal tumor burdens and robust immunity. Testing patients in this category, though, must wait until researchers are finished testing more seriously ill patient groups and have established the risks associated with the treatment. As this research so well illustrates, the development of new cancer therapies is a very complex and lengthy process.

A related gene therapy involves antigens that are found predominantly on cancer cells. During the past three to four years, scientists have made remarkable progress in identifying antigens produced by tumor cells. In addition, they have uncovered the genes that encode these tumor-associated antigens, particularly those on the most serious form of skin cancer, malignant melanoma. Now that at least some of these antigens have been described, it might be possible to develop a vaccine to prevent cancer, much like the vaccines for preventing tetanus or polio. The approach might also help treat existing tumors.

Preventive Immunization

As with the cytokine vaccines, these antigen-based cancer vaccines require gene transfer. They work best when administered to cells that are readily accessed by the immune system. For example, Philip L. Felgner of Vical in San Diego and Jon A. Wolff of the University of Wisconsin and their colleagues observed that injecting a DNA fragment coding for a foreign antigen directly into muscle triggered a potent immune response to the antigen in mice. The explanation for this reaction is simple: a bit of the foreign DNA enters the cells of the muscle or other nearby cells and directs them to produce a small amount of its protein product. Cells containing this newly synthesized foreign protein then present it to roving antibody-producing *B* cells and *T* cells. As a result, these sensitized immune components travel the body, prepared to attack tumor cells bearing the activating antigen.

The same basic strategy is revolutionizing the development of vaccines for preventing many infectious diseases. When these DNA immunizations are tested against cancer, the genes for newly identified tumor antigens are delivered directly into the body by way of vaccinia or adenovirus particles that have been rendered harmless or by such nonviral gene delivery systems as naked DNA. At present, the tests involve patients with widely spread cancer. It is clearly too late in these cases for DNA vaccines to prevent disease, but the studies should demonstrate whether the antigens can meet the essential requirement of eliciting a defensive response in the human body. Further, the studies offer a sense of whether DNA vaccines might have any merit for treating existing cancers. Given how sick many of these patients are, though, the results so far are difficult to interpret.

Yet another gene immunotherapy for cancer currently being tested in patients and in the laboratory involves antibodies. Thanks to highly variable regions on individual antibodies, these molecules are exquisitely specific. They can distinguish the slightest differences between foreign or mutated and very similar self-antigens. As it turns out, specific antibody molecules exist naturally in the outer membranes of some cancer cells—such as lymphomas that develop from *B* cells, which are committed to producing antibody molecules. Because a single lineage or clone of cells produces one specific antibody, all cancers of these cells will contain the same specific membrane molecule. This antibody then provides a unique molecular marker by which the cancer cells might be differentiated from similar but noncancerous antibody-producing cells.

Immunotherapy aside, cancers can be battled on other genetic fronts. There has been intense interest in identifying the precise DNA defects that cause cancer. Some mutations, scientists have learned, are associated with specific types of cancer. Other mutations occur in many varieties. Furthermore, there are different kinds of mutations. Some activate genes, called oncogenes, that drive uncontrolled growth in cells. Other mutations—those in so-called tumor suppressor genes—result in the loss of a normal brake on uncontrolled cell growth.

One of the most commonly mutated tumor suppressor genes in human cancer is p53, a gene whose protein product normally monitors the DNA in a cell as it divides. If the DNA is flawed, the p53 protein may halt cell division until the damage can be fixed or may induce cell suicide (apoptosis). When a normal copy of p53 is reinroduced to cancer cells in tissue culture, those cells return to a more regular growth pattern or self-destruct. Either outcome would be useful in cancer treatment, and so a great deal of effort has gone into developing methods for inserting normal p53 genes into cancers growing in the body.

There are still major roadblocks: current technologies for delivering genes to specific organs or cell populations are inefficient. In addition, there are no perfected means for extending the effects of such locally delivered genes to other areas in the body. Until physicians can do so, these therapies will help tackle only tumors at isolated sites.

Even so, animals have shown significant improvements when the p53 gene is delivered either through the bloodstream (in complexes with lipids that allow cells to take up the gene) or to tumors directly (using modified viruses to shuttle the gene into cells). An early clinical trial has reported some tumor regressions at local sites. In theory, though, there is one major limitation to using gene transfers to activate tumor suppressor genes or to neutralize oncogenes—the corrective gene must be delivered to every tumor cell. Otherwise, the unaccessed cells will continue growing uncontrollably. It is impossible to correct the genes in every tumor cell—even those in a single site—using current technology. And although additional treatments might help correct more tumor cells, repeated gene transfers using modified viruses often are not feasible; the immune system frequently recognizes the virus the second time and destroys it before it can deliver genes to tumors.

Fortunately, though, the beneficial effects of an initial injection sometimes appear to reach cells that have not been gene corrected. Indeed, several different experimental gene therapies for cancer report the appearance of a "bystander effect." This phenomenon is invoked to explain why a treatment sometimes kills a higher proportion of tumor cells than can be accounted for by the number of cells actually expressing some new gene. Researchers have reported this kind of discrepancy in some p53 gene therapy trials but cannot yet explain it: presumably, if normal p53 genes did generate a bystander effect, cancer would not develop in the first place. But the bystander effect has been seriously studied in conjunction with other treatments, such as "suicide" gene therapy, in which a gene inserted into a cancer cell renders it supersensitive to some drug that ordinarily has no anticancer effect.

In the original application of suicide gene therapy, my colleagues Edward H. Oldfield, Zvi Ram and Ken Culver and I inserted the gene for an enzyme called thymidine kinase (tk) from a herpesvirus into cancerous brain cells of patients. In cells infected with the herpes simplex virus, this enzyme can convert the otherwise nontoxic drug ganciclovir into a metabolite, or by-product, that acts as a potent viral killer. We found that this same toxic metabolite could kill dividing cancer cells; in some tumors, it killed neighboring cancer cells as well. To create this bystander effect, the toxic metabolite spread from the cell in which it was produced to its neighbors via gap junctions—channels that allow small compounds to move between cells. In the original clinical trial testing this treatment for brain tumors, about one-quarter of the patients responded. And clinicians are testing other suicide gene therapies involving different anticancer compounds, some of which are also expected to produce a bystander effect.

In various early explorations of gene therapy technology, researchers are just beginning to learn about its potential—and its limitations. As with so many other new and unexplored areas of science, some ideas will probably prove useful; many more will fall by the wayside. Ideas that are unworkable now may eventually become highly successful, when our technological capabilities increase. Even though in the future current methods for using genes for treatment will be looked back on as crude and inefficient, these methods have already offered important lessons. And they have indicated many new paths in the quest for cancer control.

Gene Therapy for AIDS

R. Michael Blase

Although about half of all clinical gene-therapy research today focuses on cancer, next largest group of studies—about 10 percent—is devoted to combating infection by the AIDS-causing human immunodeficiency virus (HIV). Experimental gene therapies for HIV aim to do one of two things: stop HIV from replicating inside infected cells, or prevent the virus from spreading to healthy ones.

The ultimate target for all these efforts would be stem cells, which develop into immune and blood cells. These cells might then be rendered resistant to HIV long before they had matured. For now, though, researchers can at best test various strategies on blood cells called monocytes and on so-called helper, or CD4, *T* cells, the immune cells most heavily ravaged by HIV. It is not difficult to isolate *T* cells from someone's blood, give them a therapeutic gene and then return them to that same person or another recipient.

Some researchers hope one day to administer therapeutic genes in vivo through delivery vehicles, or vectors, that can seek out infected or otherwise susceptible cells. The favored vectors at the moment are viruses. These viruses—which so far include adeno-associated virus and such retroviruses as HIV itself—are modified so that they are no longer pathogenic but can pass genetic information on to human cells. HIV particles altered to serve as vectors, for example, are attracted to the same CD cells as wild-type HIV particles are but do not multiply dangerously in cells. Thus, these vectors are capable of providing therapeutic genes to precisely those cells that most need them.

A range of genes are being tested for their ability to disable HIV. One gene type, containing what is called a dominant negative mutation, generates inactive versions of proteins that HIV normally makes in order to replicate. When an infected, treated cell produces these useless look-alikes, the altered proteins trip up their ordinary cousins—either by binding to them or by taking their place in molecular reactions. Clinical testing of a dominant negative mutation of the HIV gene *rev* began in 1995.

Scientists are also evaluating the merit of delivering genes that would be transcribed into short RNA strands that mimic essential viral control RNAs. The hope is that these RNA decoys might bind to HIV regulatory proteins and block them from functioning. Genes transcribed into ribozymes (catalytic RNAs) capable of degrading viral RNA might similarly interrupt HIV replication. A related idea involves delivering genes encoding proteins that are made by the host cell and that interact with HIV particles. For instance, soluble forms of the protein CD4 might bind to HIV particles extracellularly, thereby keeping them from infecting *T* cells that display CD4 molecules on their outer surface.

Scientists are also exploring for HIV treatment so-called suicide genes, which are not unlike those being tested as cancer gene therapies. Because the gene would presumably get into any cell normally invaded by the selected vector, researchers want to be sure the suicide gene will be expressed only in the subset of recipient cells that harbor HIV infection. So they plan to attach the gene to control elements that become active, and switch on the gene, only in cells that are infected by HIV.

Another design borrowed from gene therapies for cancer relies on enhancing the ability of functioning helper *T* cells to recognize infected cells and orchestrate an immune response against them. For instance, a gene coding for part of the antibody molecule that recognizes and binds to the gp 120 protein on HIV's surface can be integrated with genes encoding the molecule, or receptor, on killer *T* cells that is normally responsible for recognizing diseased cells. The chimeric receptor that results from this mix takes better notice of HIV and thus redirects the *T* cells to destroy the infected HIV cell. A Phase 1 clinical trial (looking at safety) is currently under way.

Other therapeutic genes being scrutinized give rise to antibody fragments that act within infected cells. By binding to some newly made viral protein, these intracellular antibodies, or intrabodies, prevent virus particles from being assembled. Finally, clinical trials of gene vaccinations for preventing AIDS have been approved. As in other gene vaccinations, these immunizations supply to cells patrolled by the immune system genes coding for HIV molecules that distinguish the virus as a foreign invader. The immune system then reacts to these antigens by producing antibodies that wander through the body, ready to attack any cells presenting the antigens, should they ever appear.

Suggested Reading

Barry, M. A., W. C. Lai, and S. A. Johnston. "Protection Against Myco-plasma Infection Using Expression-Library Immunization," *Nature* 377 (October 19, 1995): 632–635.

Cohen, Adam A., Jean D. Boyer, and David B. Weiner. "Modulating the Immune Response to Genetic Immunization," *FASEB Journal* 12, no. 15, (December 1998): 1611–1626.

Robinson, Harriet L. et al. "Neutralizing Antibody-Independent Contain-ment of Immunodeficiency Virus Challenges by DNA Priming and Recom-binant PoxVirus Booster Immunization," *Nature Medicine* 5, no. 5 (May 1999): 526–534.

Roman, Mark et al. "Immunostimulatory DNA Sequences Function as T Helper-I-Promoting Adjuvants," *Nature Medicine* 3, no. 8 (August 1997): 849–854.

Ulmer, J. B. et al. "Heterlogous Protection Against Influenza by Injection of DNA Encoding a Viral Protein," *Science* 259 (March 19, 1993): 1745–1749.

Chapter 6

What Can We Do with Cloning?

I, Clone

Ronald M. Green

With the first five years of this century, a team of scientists somewhere in the world will probably announce the birth of the first cloned human baby. Like Louise Brown, the first child born as the result of in vitro fertilization 21 years ago, the cloned infant will be showered with media attention. But within a few years it will be just one of hundreds or thousands of such children around the world.

It has been possible to envision such a scenario realistically only since Ian Wilmut and his colleagues at the Roslin Institute near Edinburgh, Scotland, announced in February 1997 that they had cloned a sheep named Dolly from the udder cells of a ewe. The technique used by Wilmut and his coworkers—a technology called somatic-cell nuclear transfer—will probably be the way in which the first human clone will be created.

In somatic-cell nuclear transfer, researchers take the nucleus—which contains the DNA that comprises an individual's genes—of one cell and inject it into an egg, or ovum, whose own nucleus has been removed. The resulting embryo, which will carry the nucleus donor's DNA in every one of its cells, is then implanted into the womb of a female and carried to term.

Such research on the basic processes of cell differentiation holds out the promise of dramatic new medical interventions and cures. Burn victims or those with spinal cord injuries might be provided with replacement skin or nerve tissue grown from their own body cells. The damage done by degenerative disorders such as diabetes, Parkinson's disease or Alzheimer's disease might be reversed. In the more distant future, scientists might be able to grow whole replacement organs that our bodies will not reject.

These important medical uses of cloning technology urge us to be careful in our efforts to restrict cloning research. In the immediate wake of

Dolly, politicians around the world proposed or implemented bans on human cloning. In the U.S., President Bill Clinton instituted a moratorium on federal funding for human cloning experiments, and the National Bioethics Advisory Commission urged that the ban be extended to private-sector research as well. Congress continues to study various proposals for enacting such a total ban.

In view of the still unknown physical risks that cloning might impose on the unborn child, caution is appropriate. Of the 29 early embryos created by somatic-cell nuclear transfer and implanted into various ewes by Roslin researchers, only one, Dolly, survived, suggesting that the technique currently has a high rate of embryonic and fetal loss. Dolly herself appears to be a normal three-year-old sheep—she recently gave birth to triplets following her second pregnancy. But a recent report that her telomeres—the tips of chromosomes, which tend to shrink as cells grow older—are shorter than normal for her age suggests that her life span might be reduced. This and other matters must be sorted out and substantial further animal research will need to be completed before cloning can be applied safely to humans.

Eventually animal research may indicate that human cloning can be done at no greater physical risk to the child than IVF posed when it was first introduced. One would hope that such research will be done openly in the U.S., Canada, Europe or Japan, where established government agencies exist to provide careful oversight of the implications of the studies for human subjects. Less desirably, but more probably, it might happen in clandestine fashion in some offshore laboratory where a couple desperate for a child has put their hopes in the hands of a researcher seeking instant renown.

Given the pace of events, it is possible that this researcher is already at work. For now, the technical limiting factor is the availability of a sufficient number of ripe human eggs. If Dolly is an indication, hundreds might be needed to produce only a few viable cloned embryos. Current assisted-reproduction regimens that use hormone injections to induce egg maturation produce at best only a few eggs during each female menstrual cycle. But scientists might soon resolve this problem by improving ways to store frozen eggs and by developing methods for inducing the maturation of eggs in egg follicles maintained in laboratory culture dishes.

Who First?

Once human cloning is possible, why would anyone want to have a child that way? As we consider this question, we should put aside the nightmare scenarios much talked about in the press. These include dictators using cloning to amass an army of "perfect soldiers" or wealthy egotists seeking to produce hundreds or thousands of copies of themselves. Popular films such as *Multiplicity* feed these nightmares by obscuring the fact that cloning cannot instantaneously yield a copy of an existing adult human being. What somatic-cell nuclear transfer technology produces are cloned human embryos. These require the labor- and time-intensive processes of gestation and child rearing to reach adulthood. Saddam Hussein would have to wait 20 years to realize his dream of a perfect army. And the Donald Trumps of the world would have to enlist thousands of women to be the mothers of their clones.

For all their efforts, those seeking to mass-produce children in this way, as well as others who seek an exact copy of someone else, would almost certainly be disappointed in the end. Although genes contribute to the array of abilities and limits each of us possesses, from conception forward their expression is constantly shaped by environmental factors, by the unique experiences of each individual and by purely chance factors in biological and social development. Even identical twins (natural human clones) show different physical and mental characteristics to some degree. How much more will this be true of cloned children raised at different times and in different environments from their nucleus-donor "parent"? As one wit has observed, someone trying to clone a future Adolf Hitler might instead produce a modestly talented painter.

So who is most likely to want or use human cloning? First are those individuals or couples who lack the gametes (eggs or sperm) needed for sexual reproduction. Since the birth of Louise Brown, assisted-reproduction technologies have made remarkable progress in helping infertile women and men become parents. Women with blocked or missing fallopian tubes, which carry the eggs from the ovaries to the womb, can now use in vitro fertilization to overcome the problem, and those without a functional uterus can seek the aid of a surrogate mother. A male who produces too few viable sperm cells can become a father using the technique of intracytoplasmic sperm injection, which involves inserting a single sperm or the progenitor of a sperm cell into a recipient egg.

Despite this progress, however, women who lack ovaries altogether and men whose testicles have failed to develop or have been removed must still use donor gametes if they wish to have a child, which means that the child will not carry any of their genes. Some of these individuals might prefer to use cloning technology to have a genetically related child. If a male totally lacks sperm or the testicular cells that make it, a nucleus from one of his body cells could be inserted into an egg from his mate that had had its nucleus removed. The child she would bear would be an identical twin of its father. For the couple's second child, the mother's nucleus could be used in the same procedure.

One very large category of such users of cloning might be lesbian couples. Currently if two lesbians wish to have a child, they must use donor sperm. In an era of changing laws about the rights of gamete donors, this opens their relationship to possible intervention by the sperm donor if he decides he wants to play a role in raising the child. Cloning technology avoids this problem by permitting each member of the pair to bear a child whose genes are provided by her partner. Because the egg-donor mother also supplies to each embryo a small number of mitochondria—tiny energy factories within cells that have some of their own genetic material—this approach even affords lesbian couples an approximation of sexual reproduction. (Cloning might not be used as widely by gay males, because they would need to find an egg donor and a surrogate mother.)

A second broad class of possible users of cloning technologies includes individuals or couples whose genes carry mutations that might cause serious genetic disease in their offspring. At present, if such people want a child with some genetic relationship to themselves, they can substitute donated sperm or eggs for one parent's or have each embryo analyzed genetically using preimplantation genetic diagnosis so that only those embryos shown to be free of the disease-causing gene are transferred to the mother's womb. The large number of genetic mutations contributing to some disorders and the uncertainty about which gene mutations cause some conditions limit this approach, however.

Some couples with genetic disease in their families will choose cloning as a way of avoiding what they regard as "reproductive roulette." Although the cloned child will carry the same problem genes as the parent who donates the nucleus, he or she will in all likelihood enjoy the parent's state of health and will be free of the additional risks caused by mixing both parents' genes during sexual reproduction. It is true, of course, that sex is nature's way of developing new combinations of genes that are able to resist unknown health threats in the future. Therefore, cloning should never be allowed to become so common that it reduces the overall diversity in the human gene pool. Only a relatively few couples are likely to use cloning in this way, however, and these couples will reasonably forgo the general advantages conveyed by sexual reproduction to reduce the immediate risks of passing on a genetic disease to their child.

Cloning also brings hope to families with inherited genetic diseases by opening the way to gene therapy. Such therapy—the actual correction or replacement of defective gene sequences in the embryo or the adult—is the holy grail of genetic medicine. To date, however, this research has been slowed by the inefficiency of the viruses that are now used as vectors to carry new genes into cells. By whatever means they are infused into the body, such vectors seem to reach and alter the DNA in only a frustratingly small number of cells.

Cloning promises an end run around this problem. With a large population of cells from one parent or from an embryo created from both parents' gametes, vectors could be created to convey the desired gene sequence. Scientists could determine which cells have taken up the correct sequence using fluorescent tags that cause those cells to glow. The nucleus of one of these cells could then be inserted into an egg whose own nucleus has been removed, and the "cloned" embryo could be transferred to the mother's womb. The resulting child and its descendants would thereafter carry the corrected gene in every cell of their bodies. In this way, age-old genetic maladies such as Tay-Sachs disease, cystic fibrosis, muscular dystrophy and Huntington's disease could be eliminated completely from family trees.

Cloning and Identity

Merely mentioning these beneficial uses of cloning raises difficult ethical questions. The bright hope of gene therapy is dimmed somewhat by the reawakening of eugenic fears. If we can manipulate embryos to prevent disease, why not go further and seek "enhancements" of human abilities? Greater disease resistance, strength and intelligence all beckon alluringly, but questions abound. Will we be tampering with the diversity that has been the mainstay of human survival in the past? Who will choose the alleged enhancements, and what will prevent a repetition of the terrible racist and coercive eugenic programs of the past?

Even if it proves physically safe for the resulting children, human cloning raises its own share of ethics dilemmas. Many wonder, for example, about the psychological well-being of a cloned child. What does it mean in terms of intrafamily relations for someone to be born the identical twin of his or her parent? What pressures will a cloned child experience if, from his or her birth onward, he or she is constantly being compared to an esteemed or beloved person who has already lived? The problem may be more acute if parents seek to replace a deceased child with a cloned replica. Is there, as some ethicists have argued, a "right to one's unique genotype," or genetic code—a right that cloning violates? Will cloning lead to even more serious violations of human dignity? Some fear that people may use cloning to produce a subordinate class of humans created as tissue or organ donors.

Some of these fears are less substantial than others. Existing laws and institutions should protect people produced by cloning from exploitation. Cloned humans could no more be "harvested" for their organs than people can be today. The more subtle psychological and familial harms are a worry, but they are not unique to cloning. Parents have always imposed unrealistic expectations on their children, and in the wake of widespread divorce and remarriage we have grown familiar with unusual family structures and relationships. Clearly, the initial efforts at human cloning will require good counseling for the parents and careful follow-up of the children. What is needed is caution, not necessarily prohibition.

As we think about these concerns, it is useful to keep a few things in mind. First, cloning will probably not be a widely employed reproductive technology. For many reasons, the vast majority of heterosexuals will still prefer the "old-fashioned," sexual way of producing children. No other method better expresses the loving union of a man and a woman seeking to make a baby.

Second, as we think about those who would use cloning, we would do well to remember that the single most important factor affecting the quality of a child's life is the love and devotion he or she receives from parents, not the methods or circumstances of the person's birth. Because children produced by cloning will probably be extremely wanted children, there is no reason to think that with good counseling support for their parents they will not experience the love and care they deserve.

What will life be like for the first generation of cloned children? Being at the center of scientific and popular attention will not be easy for them. They and their parents will also have to negotiate the worrisome problems created by genetic identity and unavoidable expectations. But with all these difficulties, there may also be some novel satisfactions. As cross-generational twins, a cloned child and his or her parent may experience some of the unique intimacy now shared by sibling twins. Indeed, it would not be surprising if, in the more distant future, some cloned individuals chose to perpetuate a family "tradition" by having a cloned child themselves when they decide to reproduce.

Cloning for Medicine

Ian Wilmut

In the summer of 1995 the birth of two lambs at my institution, the Roslin Institute near Edinburgh in Midlothian, Scotland, heralded what many scientists believe will be a period of revolutionary opportunities in biology and medicine. Megan and Morag, both carried to term by a surrogate mother, were not produced from the union of a sperm and an egg. Rather their genetic material came from cultured cells originally derived from a nine-day-old embryo. That made Megan and Morag genetic copies, or clones, of the embryo.

Before the arrival of the lambs, researchers had already learned how to produce sheep, cattle and other animals by genetically copying cells painstakingly isolated from early-stage embryos. Our work promised to make cloning vastly more practical, because cultured cells are relatively easy to work with. Megan and Morag proved that even though such cells are partially specialized, or differentiated, they can be genetically reprogrammed to function like those in an early embryo. Most biologists had believed that this would be impossible.

We went on to clone animals from cultured cells taken from a 26-day-old fetus and from a mature ewe. The ewe's cells gave rise to Dolly, the first mammal to be cloned from an adult. Our announcement of Dolly's birth in February 1997 attracted enormous press interest, perhaps because Dolly drew attention to the theoretical possibility of cloning humans. This is an outcome I hope never comes to pass. But the ability to make clones from cultured cells derived from easily obtained tissue should bring numerous practical benefits in animal husbandry and medical science, as well as answer critical biological questions.

How to Clone

Cloning is based on nuclear transfer, the same technique scientists have used for some years to copy animals from embryonic cells. Nuclear transfer involves the use of two cells. The recipient cell is normally an unfertilized egg taken from an animal soon after ovulation. Such eggs are poised to begin developing once they are appropriately stimulated. The donor cell is the one to be copied. A researcher working under a high-power microscope holds the recipient egg cell by suction on the end of a fine pipette and uses an extremely fine micropipette to suck out the chromosomes, sausage-shaped bodies that incorporate the cell's DNA. (At this stage, chromosomes are not enclosed in a distinct nucleus.) Then, typically, the donor cell, complete with its nucleus, is fused with the recipient egg. Some fused cells start to develop like a normal embryo and produce offspring if implanted into the uterus of a surrogate mother.

In our experiments with cultured cells, we took special measures to make the donor and recipient cells compatible. In particular, we tried to coordinate the cycles of duplication of DNA and those of the production of messenger RNA, a molecule that is copied from DNA and guides the manufacture of proteins. We chose to use donor cells whose DNA was not being duplicated at the time of the transfer. To arrange this, we worked with cells that we forced to become quiescent by reducing the concentration of nutrients in their culture medium. In addition, we delivered pulses of electric current to the egg after the transfer, to encourage the cells to fuse and to mimic the stimulation normally provided by a sperm.

After the birth of Megan and Morag demonstrated that we could produce viable offspring from embryo-derived cultures, we filed for patents and started experiments to see whether offspring could be produced from more completely differentiated cultured cells. Working in collaboration with PPL Therapeutics, also near Edinburgh, we tested fetal fibroblasts (common cells found in connective tissue) and cells taken from the udder of a ewe that was three and a half months pregnant. We selected a pregnant adult because mammary cells grow vigorously at this stage of pregnancy, indicating that they might do well in culture. Moreover, they have stable chromosomes, suggesting that they retain all their genetic information. The successful cloning of Dolly from the mammary-derived culture and of other lambs from the cultured fibroblasts showed that the Roslin protocol was robust and repeatable.

All the cloned offspring in our experiments looked, as expected, like the breed of sheep that donated the originating nucleus, rather than like their surrogate mothers or the egg donors. Genetic tests prove beyond doubt that Dolly is indeed a clone of an adult. It is most likely that she was derived from a fully differentiated mammary cell, although it is impossible to be certain because the culture also contained some less differentiated cells found in small numbers in the mammary gland. Other laboratories have since used an essentially similar technique to create healthy clones of cattle and mice from cultured cells, including ones from nonpregnant animals.

Although cloning by nuclear transfer is repeatable, it has limitations. Some cloned cattle and sheep are unusually large, but this effect has also been seen when embryos are simply cultured before gestation. Perhaps more important, nuclear transfer is not yet efficient. John B. Gurdon, now at the University of Cambridge, found in nuclear-transfer experiments with frogs more than 30 years ago that the number of embryos surviving to become tadpoles was smaller when donor cells were taken from animals at a more advanced developmental stage. Our first results with mammals showed a similar pattern. All the cloning studies described so far show a consistent pattern of deaths during embryonic and fetal development, with laboratories reporting only one to two percent of embryos surviving to become live offspring. Sadly, even some clones that survive through birth die shortly afterward.

Clones with a Difference

The cause of these losses remains unknown, but it may reflect the complexity of the genetic reprogramming needed if a healthy offspring is to be born. If even one gene inappropriately expresses or fails to express a crucial protein at a sensitive point, the result might be fatal. Yet reprogramming might involve regulating thousands of genes in a process that could involve some randomness. Technical improvements, such as the use of different donor cells, might reduce the toll.

The ability to produce offspring from cultured cells opens up relatively easy ways to make genetically modified, or transgenic, animals. Such animals are important for research and can produce medically valuable human proteins.

The standard technique for making transgenic animals is painfully slow and inefficient. It entails microinjecting a genetic construct—a DNA sequence incorporating a desired gene—into a large number of fertilized eggs. A few of them take up the introduced DNA so that the resulting offspring express it. These animals are then bred to pass on the construct [see "Transgenic Livestock as Drug Factories," by William H. Velander, Henryk Lubon and William N. Drohan; *Scientific American*, January 1997].

In contrast, a simple chemical treatment can persuade cultured cells to take up a DNA construct. If these cells are then used as donors for nuclear transfer, the resulting cloned offspring will all carry the construct. The Roslin Institute and PPL Therapeutics have already used this approach to produce transgenic animals more efficiently than is possible with micro-injection.

We have incorporated into sheep the gene for human factor IX, a blood-clotting protein used to treat hemophilia B. In this experiment we transferred an antibiotic-resistance gene to the donor cells along with the factor IX gene, so that by adding a toxic dose of the antibiotic neomycin to the culture, we could kill cells that had failed to take up the added DNA. Yet despite this genetic disruption, the proportion of embryos that developed to term after nuclear transfer was in line with our previous results.

The first transgenic sheep produced this way, Dolly, was born in the summer of 1997. Dolly and other transgenic clones secrete the human protein in their milk. These observations suggest that once techniques for the retrieval of egg cells in different species have been perfected, cloning will make it possible to introduce precise genetic changes into any mammal and to create multiple individuals bearing the alteration.

Cultures of mammary gland cells might have a particular advantage as donor material. Until recently, the only practical way to assess whether a DNA construct would cause a protein to be secreted in milk was to transfer it into female mice, then test their milk. It should be possible, however, to test mammary cells in culture directly. That will speed up the process of finding good constructs and cells that have incorporated them so as to give efficient secretion of the protein.

Cloning offers many other possibilities. One is the generation of genetically modified animal organs that are suitable for transplantation into humans. At present, thousands of patients die every year before a replacement heart, liver or kidney becomes available. A normal pig organ would be rapidly destroyed by a "hyper-acute" immune reaction if transplanted into a human. This reaction is triggered by proteins on the pig cells that have been modified by an enzyme called alpha-galactosyl transferase. It stands to reason, then, that an organ from a pig that has been genetically altered so that it lacks this enzyme might be well tolerated if doctors gave the recipient drugs to suppress other, less extreme immune reactions.

Another promising area is the rapid production of large animals carrying genetic defects that mimic human illnesses, such as cystic fibrosis. Although mice have provided some information, mice and humans have very different genes for cystic fibrosis. Sheep are expected to be more valuable for research into this condition, because their lungs resemble those of humans. Moreover, because sheep live for years, scientists can evaluate their long-term responses to treatments.

Creating animals with genetic defects raises challenging ethical questions. But it seems clear that society does in the main support research on animals, provided that the illnesses being studied are serious ones and that efforts are made to avoid unnecessary suffering.

The power to make animals with a precisely engineered genetic constitution could also be employed more directly in cell-based therapies for important illnesses, including Parkinson's disease, diabetes and muscular dystrophy. None of these conditions currently has any fully effective treatment. In each, some pathological process damages specific cell populations, which are unable to repair or replace themselves. Several novel approaches are now being explored that would provide new cells—ones taken from the patient and cultured, donated by other humans or taken from animals.

To be useful, transferred cells must be incapable of transmitting new disease and must match the patient's physiological need closely. Any immune response they produce must be manageable. Cloned animals with precise genetic modifications that minimize the human immune response might constitute a plentiful supply of suitable cells. Animals might even produce cells with special properties, although any modifications would risk a stronger immune reaction.

Cloning could also be a way to produce herds of cattle that lack the prion protein gene. This gene makes cattle susceptible to infection with prions, agents that cause bovine spongiform encephalitis (BSE), or mad cow disease. Because many medicines contain gelatin or other products derived from cattle, health officials are concerned that prions from infected animals could infect patients. Cloning could create herds that, lacking the prion protein gene, would be a source of ingredients for certifiable prion-free medicines.

The technique might in addition curtail the transmission of genetic disease. Many scientists are now working on therapies that would supplement or replace defective genes in cells, but even successfully treated patients will still pass on defective genes to their offspring. If a couple was willing to produce an embryo that could be treated by advanced forms of gene therapy, nuclei from modified embryonic cells could be transferred to eggs to create children who would be entirely free of a given disease.

Some of the most ambitious medical projects now being considered envision the production of universal human donor cells. Scientists know how to isolate from very early mouse embryos undifferentiated stem cells, which can contribute to all the different tissues of the adult. Equivalent cells can be obtained for some other species, and humans are probably no exception. Scientists are learning how to differentiate stem cells in culture, so it may be possible to manufacture cells to repair or replace tissue damaged by illness.

Making Human Stem Cells

Stem cells matched to an individual patient could be made by creating an embryo by nuclear transfer just for that purpose, using one of the patient's cells as the donor and a human egg as the recipient. The embryo would be allowed to develop only to the stage needed to separate and culture stem cells from it. At that point, an embryo has only a few hundred cells, and they have not started to differentiate. In particular, the nervous system has not begun to develop, so the embryo has no means of feeling pain or sensing the environment. Embryo-derived cells might be used to treat a variety of serious diseases caused by damage to cells, perhaps including AIDS as well as Parkinson's, muscular dystrophy and diabetes.

Scenarios that involve growing human embryos for their cells are deeply disturbing to some people, because embryos have the potential to become people. The views of those who consider life sacred from conception should be respected, but I suggest a contrasting view. The embryo is a cluster of cells that does not become a sentient being until much later in development, so it is not yet a person. In the U.K., the Human Genetics Advisory Commission has initiated a major public consultation to assess attitudes toward this use of cloning.

Creating an embryo to treat a specific patient is likely to be an expensive proposition, so it might be more practical to establish permanent, stable human embryonic stem-cell lines from cloned embryos. Cells could then be differentiated as needed. Implanted cells derived this way would not be genetically perfect matches, but the immune reaction would probably be controllable. In the longer term, scientists might be able to develop methods for manufacturing genetically matched stem cells for a patient by "dedifferentiating" them directly, without having to utilize an embryo to do it.

Several commentators and scientists have suggested that it might in some cases be ethically acceptable to clone existing people. One scenario envisages generating a replacement for a dying relative. All such possibilities, however, raise the concern that the clone would be treated as less than a complete individual, because he or she would likely be subjected to limitations and expectations based on the family's knowledge of the genetic "twin." Those expectations might be false, because human personality is only partly determined by genes. The clone of an extrovert could have a quite different demeanor. Clones of athletes, movie stars, entrepreneurs or scientists might well choose different careers because of chance events in early life.

Some pontificators have also put forward the notion that couples in which one member is infertile might choose to make a copy of one or the other partner. But society ought to be concerned that a couple might not treat naturally a child who is a copy of just one of them. Because other methods are available for the treatment of all known types of infertility, conventional therapeutic avenues seem more appropriate. None of the suggested uses of cloning for making copies of existing people is ethically acceptable to my way of thinking, because they are not in the interests of the resulting child. It should go without saying that I strongly oppose allowing cloned human embryos to develop so that they can be tissue donors.

It nonetheless seems clear that cloning from cultured cells will offer important medical opportunities. Predictions about new technologies are often wrong: societal attitudes change; unexpected developments occur. Time will tell. But biomedical researchers probing the potential of cloning now have a full agenda.

Is Quiescence the Key to Cloning?

Ian Wilmut

All the cells that we used as donors for our nuclear-transfer experiments were quiescent—that is, they were not making messenger RNA. Most cells spend much of their life cycle copying DNA sequences into messenger RNA, which guides the production of proteins. We chose to experiment with quiescent cells because they are easy to maintain for days in a uniform state. But Keith H. S. Campbell of our team recognized that they might be particularly suitable for cloning.

He conjectured that for a nuclear transfer to be successful, the natural production of RNA in the donor nucleus must first be inhibited. The reason is that cells in a very early stage embryo are controlled by proteins and RNA made in the precursor of the parent egg cell. Only about three days after fertilization does the embryo start making its own RNA. Because an egg cell's own chromosomes would normally not be making RNA, nuclei from quiescent cells may have a better chance of developing after transfer.

A related possibility is that the chromosomes in quiescent nuclei may be in an especially favorable physical state. We think regulatory molecules in the recipient egg act on the transferred nucleus to reprogram it. Although we do not know what these molecules are, the chromosomes of a quiescent cell may be more accessible to them.

Suggested Reading

Campbell, K. H. S, P. Loi, P. J. Otaegui, and I. Wilmut. "Cell Cycle Coordination in Embryo Cloning by Nuclear Transfer, *Reviews of Reproduction* 1, no. 1 (January 1996): 40–46.

Campbell, K. H. S., J. McWhir, W. A. Ritchie, and I. Wilmut. "Sheep Cloned by Nuclear Transfer from a Cultured Cell Line," *Nature* 385 (February 27, 1997): 810–813.

Di Berardino, Marie A. *Genomic Potential of Differentiated Cells.* New York: Columbia University Press, 1997.

Schnieke, E. et al. "Human Factor IX Transgenic Sheep Produced by Transfer of Nuclei from Transfected Fetal Fibroblasts," *Science* 278 (December 19, 1997): 2130–2133.

Chapter 7

Extreme Biology

Biochemist Baruch S. Blumberg: The Search for Extreme Life

Julie Wakefield

The relentless heat cooks the Badwater region of California's Death Valley so thoroughly that some expanses are textured like dry serpent skin. At some 284 feet below sea level—North America's lowest point—it is perhaps the hottest place on the surface of the earth: the temperature once peaked at a record 53.01 degrees Celsius (127.4 degrees Fahrenheit). Out here, blood-pumping mammals are scarce. It may seem unfitting to find a Nobel Prize winner, renowned for hepatitis B work, in this scorching pit. But Baruch S. Blumberg's latest challenge takes him beyond human subjects. As the first director of the National Aeronautics and Space Administration's Astrobiology Institute (NAI), he is searching for extreme life-forms, the kind the space agency aims to someday find on other worlds.

"I always liked the idea of doing fieldwork, exploring, going out and finding new things," Blumberg says back at NAI headquarters, which is nestled near Silicon Valley at the NASA Ames Research Center at Moffett Field. Out of his desert garb, the outdoors-loving Blumberg looks a good decade younger than his 75 years. At the job only since September 1999, Blumberg is trying to marshal gaggles of astronomers, chemists, ecologists, geologists, biologists, physicists and even zoologists. He is convinced that advances in molecular biology, space exploration and other endeavors make timely the reexamination of such age-old issues as the origins of life and its possible existence elsewhere.

"Technology is available to decipher the intricacies of this cause-and-effect chain" that wasn't available even five years ago, Blumberg notes, citing in particular advances achieved through the Human Genome Project.

The 1996 announcement of potential fossilized life in a Martian meteorite known as ALH84001 boosted enthusiasm worldwide. Even Congress, which had quashed NASA's search for extraterrestrial intelligence (SETI) program in 1993, became receptive. On sabbatical at Stanford University in 1998, Blumberg, along with scores of others, helped to craft NASA's Astrobiology Roadmap during a series of workshops. It defined the role for the new institute.

"With NASA's Astrobiology Institute we are witnessing not just a shift in scientific paradigm but, more important, a shift in cultural acceptability among scientists," says extrasolar planet hunter Geoffrey W. Marcy of San Francisco State University. Already Blumberg's institute is becoming "the intellectual basis for a broad range of NASA missions," says NASA administrator Daniel S Goldin. Goldin hopes to raise the NAI's budget from about $15 million to $100 million within five years. The NAI now comprises some 430 astrobiologists at 11 universities and research institutions.

Although the institute is lending new credibility to the search for extraterrestrial life, X-Files fans needn't hold their breath. Unlike the now privately funded SETI program, which focuses on radio transmissions and other hallmarks of presumably sentient beings, the NAI is targeting microorganisms and other, even more primitive evidence of lifelike matter. Specifically, the NAI is looking for life in hostile environments—in deserts, volcanoes and ice caps; down thousands of meters below Earth's surface or into the ocean; and on Mars, Jupiter's moon Europa, Saturn's satellite Titan, even planets beyond the solar system.

For now at least, extremophiles on Earth offer the most probable model for testing the hypothesis that life exists elsewhere. NAI researchers hope to use genomic databases of key microorganisms to link evolutionary sequences with geochemical and paleontological events. Another desire is to launch DNA microprobes on board miniature spacecraft to search for signs of life. Answers, if they ever come, may take many decades.

Blumberg believes his past biochemical work gives him intimate insights into life-forms, whether of this world or not. "One of the things about doing medicine and medical research is that you really get a kind of feeling for the organism that you work with," he observes. Hence, profound questions of life "are coming directly and indirectly into your thinking."

As a child in a tight-knit immigrant community in Brooklyn, New York, Blumberg checked out book after library book on the reigning explorers. "Amundsen, Peary, Scott, Shackleton, Rae, Nansen were common names in my circle of friends," he recalls. "I believe this had an effect on my seeing science as discovery. My interest in fieldwork also fed into this." To this day he collects books on early travel and Arctic expeditions.

After graduating from Far Rockaway High School in 1943, he enlisted in the Naval Reserves and secured a physics degree at Union College in Schenectady, New York. At age 21 he made captain of a small U.S. Navy ship. "It is a great sensation to plot a course, take a few sights, do some dead reckoning, and end up more or less where you had predicted. It gives one confidence in the power of applied mathematics and the effectiveness of rational solutions." Captaining that crew 24 hours a day instilled an unshakable confidence in him. "I assumed that I would have leadership roles in whatever I did," he says.

In 1946, thanks to the G.I. Bill, Blumberg started graduate school in mathematics at Columbia University, only to transfer a year later to the medical school at the behest of his attorney father. For his medical internship and residency, Blumberg picked the crowded, understaffed wards of New York City's Bellevue Hospital, where the poor and chronically ill were typically sent. "And this was before health insurance," he emphasizes. Bellevue taught Blumberg a new definition of responsibility: "The fact that you've got to do it—if you don't do it, nobody else will."

Equipped with an M.D., he decided to pursue his own longing to be a scientist and went in 1955 to the University of Oxford, where he began his doctorate in biochemistry under Alexander G. Ogston. At the time, Oxbridge was buzzing with excitement over Watson and Crick's discovery of the DNA double helix. Blumberg himself had become intrigued with inherited genetic variations a few years earlier. In 1950 he had gone to a desolate mining-town hospital in Suriname in South America, where, besides witnessing the devastation caused by infectious diseases, he observed large differences in susceptibility to the elephantiasis parasite among diverse immigrant workers. A 1957 field trip to West Africa formally launched his study of such genetic variations, called polymorphisms, which he would continue at the National Institutes of Health.

Blumberg collected data on the distribution of polymorphisms. Initially, he culled blood for clues to disease resistance. To find possible variants, he and his colleagues relied on the natural immune response to compare blood proteins from frequently transfused patients, mainly hemophiliacs. From antibodies in the patients' bloodstream, they could derive foreign antigens. In 1963 Blumberg's team isolated a peculiar variant and dubbed it "Australian antigen." Common among Australian Aborigines, Micronesians, Vietnamese and Taiwanese, the blood protein was rare among Westerners. The team, however, observed it in leukemia patients in the U.S., who also were receiving transfusions. The researchers set off exploring whether the unusual antigen played a role in susceptibility to leukemia.

Instead of an inherited immune factor, the curious surface antigen proved to be part of the then mysterious hepatitis B virus. "His discovery of Australian antigen was the Rosetta stone for unraveling the nature of the hepatitis viruses," comments Robert H. Purcell, head of the NIH's hepatitis lab.

This key finding enabled researchers to develop the first blood test to screen for the virus, thus protecting blood supplies. In 1969 Blumberg and microbiologist Irving Millman patented a strategy to develop a hepatitis B vaccine. Their novel approach relied on purifying from the virus those very same surface antigen particles—which by good fortune proved not only to produce protective antibodies but to be noninfectious. For advancing understanding of the mechanisms of infectious diseases, Blumberg shared the 1976 Nobel Prize for Physiology or Medicine.

A commercial vaccine based on Blumberg's method, now made using recombinant DNA techniques, has saved tens of millions of lives, according to World Health Organization estimates. Blumberg remains optimistic that hepatitis B can someday be eradicated, but today the virus continues to kill more than a million people a year, including 5,000 in the U.S.

When not working, the Nobelist prefers to birdwatch or kayak or even shovel manure on a cattle farm he owns with friends in western Maryland. "That kind of manual labor is an antidote to too much thinking," he says. In Death Valley, Blumberg and other researchers, led by Christopher McKay of NASA Ames, used syringes to extract heat-loving microbes for DNA analysis back at the lab. Blumberg plans to accompany researchers on other field trips to collect extremophiles, perhaps in Mongolia's Gobi Desert or in Antarctica. Tests of new robots for planetary exploration might even send him to the Canadian Arctic.

Besides guiding and inspiring his researchers, Blumberg wants to take advantage of powerful computers to model how life might evolve elsewhere. "Astrobiology lends itself to iterated induction-deduction exercises, as well as theory and model construction," Blumberg explains. He notes wryly that in this field "there's a high probability you will reject the model." Just the same, he and his followers hope the conditions that allow life to flourish on Earth exist elsewhere in the Milky Way and beyond. "It could happen," Blumberg says. "In any case, you have to go and look."

Car Parts from Chickens

Diane Martindale

Nearly a decade ago poultry-processing plants around the nation asked researchers at the Department of Agriculture to solve a big environmental problem: find a more efficient way to dispose of the four billion pounds of chicken feathers produced annually in the U.S. What they were expecting was a method by which the feathers could be made more biodegradable after burial. But Walter Schmidt, a chemist at the Agricultural Research Service (ARS) in Beltsville, Maryland, went a step further to develop a recycling technology that will soon bring feathers into everyday life disguised as plastic and paper products.

Currently poultry farmers mix water with leftover feathers in large pressure cookers to make low-grade feedstuff for chickens and cattle—a venture that is generally not profitable. But converting feathers into value-added products required more than just a little steam. Schmidt and his colleagues developed an efficient mechanical method to separate the more valuable barb fibers (plumage) from the less useful central chaff, or quill. Though softer, the keratin fibers in the barbs are stronger and less brittle than those in the quill and therefore have a much broader range of applications.

The key to easy separation lay in the fact that the quills are bulkier and heavier. The feathers, dried and sterilized, are shredded and fed into a cylindrical device consisting of an outer and inner tube. The feathers are sucked through the central channel, and the quills are drawn off at the bottom, but thanks to air turbulence, the barbs float back up between the sides of the tubes.

Once separated, barb fibers can be used in many ways. Schmidt and his collaborators have made diaper filler, paper towels and water filters out of them. The ground fibers have been used in plastics, in pulp to make paper and in combination with synthetic and natural fibers to make textiles. And the fibers are good for more chemically complex applications as well. For instance, by mixing the fiber powder with a reducing agent and placing the slurry in a hydraulic press, Attila Pavlath, a scientist for the ARS in California, has created polymer films. "The reducing agent acts like a hairdresser's perm solution to relax the protein bonds of the keratin, allowing us to mold the fiber into thin sheets of plastic," Pavlath explains. This polymer may first show up as biodegradable candy wrappers (similar to cellophane) and six-pack can-holders.

The powder can also replace additives, such as nonrecyclable fiberglass, that are used to strengthen plastic. Combined with polyethylene, the barbs can produce a more rigid plastic suitable for dog-food bowls and automobile interior parts, including the dashboard.

The quill portion doesn't have to go to waste, either. David Emery of Featherfiber Corporation in Nixa, Missouri, has developed a process to make high-grade quill protein that is 90 percent digestible (typical quill meal is only 50 percent digestible), Emery says. The company has licensed Schmidt's patents and has just completed a pilot plant to produce feather fiber.

Farm animals may not be the only ones to benefit from a quill meal. Carlo Licata of MaXim LLC in Pasadena, California, believes that the quill portion is an excellent dietary supplement for humans. "That's because the keratin protein is very absorbent," Licata indicates, "and can retain nutrients for a longer period"—something like Metamucil, only better.

All this and more from chicken feathers without breaking the farm. "A typical farm produces ten thousand pounds of feathers per hour, which is enough to meet the needs of one plastic-producing plant," Schmidt remarks. If all the feathers in the U.S. were processed, more than five billion pounds of plastic products could be made.

Feather-derived plastics are just one of several nonpetroleum-based "green plastics" that have surfaced in the past year. Cargill-Dow Polymers in Minnetonka, Minnesota, recently announced production of a new kind of natural plastic made from polylactic acid, a compound derived from corn. Monsanto, maker of genetically modified plants, reported in October 1999 that it had fabricated a plant capable of producing biodegradable plastic of a type known as polyhydroxyalkanoate.

But the consequences of producing greener plastics are often overlooked, according to Tillman Gerngross, a biochemical engineer at Dartmouth College. "People too readily accept the premise that renewable equals environmentally good. It does not necessarily add up." If you have to use huge amounts of coal to make the plastics, then you are harming the environment just the same, he points out. And feather plastics are often only partially biodegradable. Still, Gerngross agrees that a move toward sustainable resources is desirable. That should prevent researchers like Walter Schmidt from chickening out too soon.

Building a Brainier Mouse

Joe Z. Tsien

When I decided to become a scientist, never in my wildest dreams did I imagine that my work would provide fodder for CBS's *Late Show with David Letterman*. But in September 1999, after my colleagues and I announced that we had doctored the genes of some mice to enhance their learning and memory skills, I turned on my television to find that my creations were the topic of one of Letterman's infamous Top Ten Lists. As I watched, the comedian counted down his roster of the Top Ten Term Paper Topics Written by Genius Mice. (My personal favorites are "Our Pearl Harbor: The Day Glue Traps Were Invented" and "Outsmarting the Mousetrap: Just Take the Cheese Off Really, Really Fast.")

My furry research subjects had become overnight celebrities. I received mail by the bagful and was forwarded dozens of jokes in which "smart" mice outwitted duller humans and their feeble traps. It seemed that the idea of a more intelligent mouse was something that everyone could identify with and find humorous.

But my coworkers and I did not set out merely to challenge the inventiveness of mousetrap manufacturers. Our research was part of a decades-long line of inquiry into exactly what happens in the brain during learning and what memories are made of. By generating the smart mice—a strain that we dubbed *Doogie* after the boy genius on the TV show *Doogie Howser, M.D.*—we validated a 50-year-old theory about the mechanisms of learning and memory and illustrated the central role of a particular molecule in the process of memory formation. That molecule could one day serve as a possible target for drugs to treat brain disorders such as Alzheimer's disease or even, perhaps, to boost learning and memory capacity in normal people.

Understanding the molecular basis of learning and memory is so important because what we learn and what we remember determine largely who we are. Memory, not merely facial and physical appearance, defines an individual, as everyone who has known someone with Alzheimer's disease understands all too well. Furthermore, learning and memory extend beyond the individual and transmit our culture and civilization over generations. They are major forces in driving behavioral, cultural and social evolution.

The ABCs of Learning and Memory

The human brain has approximately 100 billion nerve cells, or neurons, that are linked in networks to give rise to a variety of mental and cognitive attributes, such as memory, intelligence, emotion and personality. The foundations for understanding the molecular and genetic mechanisms of learning and memory were laid in 1949, when Canadian psychologist Donald O. Hebb came up with a simple yet profound idea to explain how memory is represented and stored in the brain. In what is now known as Hebb's learning rule, he proposed that a memory is produced when two connected neurons are active simultaneously in a way that somehow strengthens the synapse, the site where the two nerve cells touch each other. At a synapse, information in the form of chemicals called neurotransmitters flows from the so-called presynaptic cell to one dubbed the postsynaptic cell.

In 1973 Timothy V. P. Bliss and Terje Lømo, working in Per Andersen's laboratory at the University of Oslo, discovered an experimental model with the hallmark features of Hebb's theory. They found that nerve cells in a sea horse–shaped region of the brain, appropriately called the hippocampus (from the Greek for "horse-headed sea monster"), become more tightly linked when stimulated by a series of high-frequency electrical pulses. The increase in synaptic strength—a phenomenon known as long-term potentiation (LTP)—can last for hours, days or even weeks. The fact that LTP is found in the hippocampus is particularly fascinating because the hippocampus is a crucial brain structure for memory formation in both humans and animals.

Later studies by Mark F. Bear of the Howard Hughes Medical Institute at Brown University and other scientists showed that applying a low-frequency stimulation to the same hippocampal pathway produces a long-lasting *decrease* in the strength of the connections there. The reduction is also long-lasting and is known as long-term depression (LTD), although it apparently has nothing to do with clinical depression.

The strengthening and weakening of synaptic connections through LTP- and LTD-like processes have become the leading candidate mechanisms for storing and erasing learned information in the brain. We now know that LTP and LTD come in many different forms. The phenomena also occur in many brain regions besides the hippocampus, including the neocortex—the "gray matter"—and the amygdala, a structure involved in emotion.

What is the molecular machinery controlling these forms of synaptic changes, or plasticity? Studies in the 1980s and 1990s by Graham L. Collingridge of the University of Bristol in England, Roger A. Nicoll of the University of California at San Francisco, Robert C. Malenka of Stanford University, Gary S. Lynch of the University of California at Irvine and other researchers have found that the changes depend on a single type of molecule. The researchers demonstrated that the induction of the major forms of LTP and LTD requires the activation of so-called NMDA receptors, which sit on the cell membranes of postsynaptic neurons.

NMDA receptors are really minuscule pores that most scientists think are made up of four protein subunits that control the entry of calcium ions into neurons. (The name of the receptors derives from N-methyl-D-aspartate, an artificial chemical that happens to bind to them.) They are perfect candidates for implementing the synaptic changes of Hebb's learning rule because they require two separate signals to open—the binding of the neurotransmitter glutamate and an electrical change called membrane depolarization. Accordingly, they are the ideal molecular switches to function as "coincidence detectors" to help the brain associate two events.

Although LTP and LTD had been shown to depend on NMDA receptors, linking LTP- and LTD-like processes to learning and memory turned out to be much more difficult than scientists originally thought. Richard G. M. Morris of the University of Edinburgh and his colleagues have observed that rats whose brains have been infused with drugs that block the NMDA receptor cannot learn how to negotiate a test called a Morris water maze as well as other rats. The finding is largely consistent with the prediction for the role of LTP in learning and memory. The drugs often produce sensory-motor and behavioral disturbances, however, indicating the delicate line between drug efficacy and toxicity.

Four years ago, while I was working in Susumu Tonegawa's laboratory at the Massachusetts Institute of Technology, I went one step further and developed a new genetic technique to study the NMDA receptor in learning and memory. The technique was a refinement of the method for creating so-called knockout mice—mice in which one gene has been selectively inactivated, or "knocked out." Traditional knockout mice lack a particular gene in every cell and tissue. By studying the health and behavior of such animals, scientists can deduce the function of the gene.

But many types of knockout mice die at or before birth because the genes they lack are required for normal development. The genes encoding the various subunits of the NMDA receptors turned out to be similarly essential: regular NMDA-receptor knockout mice died as pups. So I devised a way to delete a subunit of the NMDA receptor in only a specific region of the brain.

Scoring a Knockout

Using the new technique, I engineered mice that lacked a critical part of the NMDA receptor termed the NR1 subunit in a part of their hippocampus known as the CA1 region. It was fortunate that we knocked out the gene in the CA1 region because that is where most LTP and LTD studies have been conducted and because people with brain damage to that area have memory deficits. In collaboration with Matthew A. Wilson, Patricio T. Huerta, Thomas J. McHugh and Kenneth I. Blum of M.I.T., I found that the knockout mice have lost the capacity to change the strength of the neuronal connections in the CA1 regions of their brains. These mice exhibit abnormal spatial representation and have poor spatial memory: they cannot remember their way around a water maze. More recent studies in my own laboratory at Princeton University have revealed that the mice also show impairments in several other, nonspatial memory tasks.

Although these experiments supported the hypothesis that the NMDA receptors are crucial for memory, they were not fully conclusive. The drugs used to block the receptors could have exerted their effects through other molecules in addition to NMDA receptors, for example. And the memory deficits of the knockout mice might have been caused by another, unexpected abnormality independent of the LTP/LTD deficits.

To address these concerns, a couple of years ago I decided to try to increase the function of NMDA receptors in mice to see whether such an alteration improved the animals' learning and memory. If it did, that result—combined with the previous ones—would tell us that the NMDA receptor truly is a central player in memory processes.

This time I focused on different parts of the NMDA receptor, the NR2A and NR2B subunits. Scientists have known that the NMDA receptors of animals as diverse as birds, rodents and primates remain open longer in younger individuals than in adults. Some researchers, including my colleagues and me, have speculated that the difference might account for the fact that young animals are usually able to learn more readily—and remember what they have learned longer—than their older counterparts.

As individuals mature, they begin to switch from making NMDA receptors that contain NR2B subunits to those that include NR2A subunits. Laboratory studies have shown that receptors with NR2B subunits stay open longer than those with NR2A. I reasoned that the age-related switch could explain why adults can find it harder to learn new information.

So I took a copy of the gene that directs the production of NR2B and linked it to a special piece of DNA that served as an on switch to specifically increase the gene's ability to make the protein in the adult brain. I injected this gene into fertilized mouse eggs, where it was incorporated into the chromosomes and produced genetically modified mice carrying the extra copy of the NR2B gene.

Working in collaboration with Guosong Liu of M.I.T. and Min Zhuo of Washington University, my colleagues and I found that NMDA receptors from the genetically engineered mice could remain open for roughly 230 milliseconds, almost twice as long as those of normal mice. We also determined that neurons in the hippocampi of the adult mice were capable of making stronger synaptic connections than those of normal mice of the same age. Indeed, their connections resembled those in juvenile mice.

What Smart Mice Can Do

Next, Ya-Ping Tang and other members of my laboratory set about evaluating the learning and memory skills of the mice that we had named *Doogie*. First, we tested one of the most basic aspects of memory, the ability to recognize an object. We placed *Doogie* mice into an open box and allowed them to explore two objects for five minutes. Several days later we replaced one object with a new one and returned the mice to the box. The genetically modified mice remembered the old object and devoted their time to exploring the new one. Normal mice, however, spent an equal amount of time exploring both objects, indicating that the old object was no more familiar to them than the new. By repeating the test at different intervals, we found that the genetically modified mice remembered objects four to five times longer than their normal counterparts did.

In the second round of tests, Tang and I examined the ability of the mice to learn to associate a mild shock to their paws with being in a particular type of chamber or hearing a certain tone. We found that the *Doogie* mice were more likely to "freeze"—an indication that they remembered fear—than were normal mice when we returned the animals to the chamber or played them the tone several days later. These tests suggested to us that the *Doogie* mice had better memory. But were they also faster learners?

Learning and memory represent different stages of the same gradual and continuous process whose steps are often not easy to distinguish. Without memory, one cannot measure learning; without learning, no memory exists to be assessed. To determine whether the genetic alteration of the *Doogie* mice helped them to learn, we employed a classic behavioral experimental paradigm known as fear-extinction learning.

In the fear-extinction test, we conditioned the mice as we did before in a shock chamber, then placed the animals back into the fear-causing environment—but without the paw shocks—again and again. Most animals take five repetitions or so to unlearn the link between being in the shock chamber and receiving a shock. The *Doogie* mice learned to be unafraid after only two repetitions. They also learned not to fear the tone faster than the normal mice.

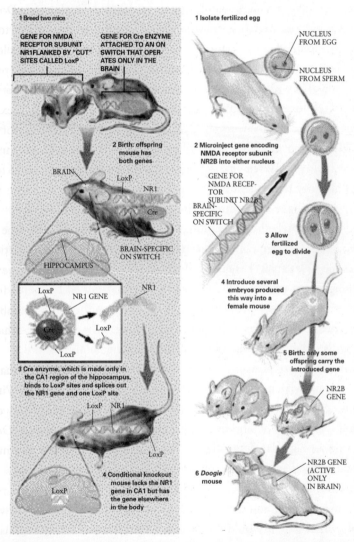

HOW TO MAKE A DUMB MOUSE

Remove part of a key receptor from its brain

1 Breed two mice

GENE FOR NMDA RECEPTOR SUBUNIT NR1 FLANKED BY "CUT" SITES CALLED LoxP

GENE FOR Cre ENZYME ATTACHED TO AN ON SWITCH THAT OPERATES ONLY IN THE BRAIN

2 Birth: offspring mouse has both genes

BRAIN

LoxP

NR1

Cre

BRAIN-SPECIFIC ON SWITCH

HIPPOCAMPUS

LoxP   NR1 GENE   NR1

Cre   LoxP

LoxP

3 Cre enzyme, which is made only in the CA1 region of the hippocampus, binds to LoxP sites and splices out the NR1 gene and one LoxP site

LoxP   NR1

LoxP

4 Conditional knockout mouse lacks the NR1 gene in CA1 but has the gene elsewhere in the body

LoxP

HOW TO MAKE A SMART MOUSE

Add an extra copy of part of a key receptor to its brain

1 Isolate fertilized egg

NUCLEUS FROM EGG

NUCLEUS FROM SPERM

2 Microinject gene encoding NMDA receptor subunit NR2B into either nucleus

GENE FOR NMDA RECEPTOR SUBUNIT NR2B

BRAIN-SPECIFIC ON SWITCH

3 Allow fertilized egg to divide

4 Introduce several embryos produced this way into a female mouse

5 Birth: only some offspring carry the introduced gene

NR2B GENE

6 *Doogie* mouse

NR2B GENE (ACTIVE ONLY IN BRAIN)

Making dumb and smart mice involves tampering with a protein called the NMDA receptor that is important for learning and memory. But the NMDA receptor plays crucial roles elsewhere in the body, so the author and his colleagues used snippets of DNA (on switches in the diagram) to manipulate the genes for various subunits of the receptor *only* in the brain. The smart, or *Doogie*, mice have extra subunits in their brains; the dumb, or conditional knockout, mice lack a different NMDA receptor subunit in their brains.

The last behavioral test was the Morris water maze, in which the mice were required to use visual cues on a laboratory wall to find the location of a submerged platform hidden in a pool of milky water. This slightly more complicated task involves many cognitive factors, including analytical skills, learning and memory, and the ability to form strategies. Again, the genetically modified mice performed better than their normal counterparts.

Our experiments with *Doogie* mice clearly bore out the predictions of Hebb's rule. They also suggested that the NMDA receptor is a molecular master switch for many forms of learning and memory.

Although our experiments showed the central role of NMDA receptors in a variety of learning and memory processes, it is probably not the only molecule involved. We can expect many molecules that play a role in learning and memory to be identified in the coming years.

Everyone I have encountered since the publication of our results has wanted to know whether the findings mean we will soon be able to genetically engineer smarter children or devise pills that will make everyone a genius. The short answer is no—and would we even want to?

Intelligence is traditionally defined in dictionaries and by many experimental biologists as "problem-solving ability." Although learning and memory are integral parts of intelligence, intelligence is a complex trait that also involves many other factors, such as reasoning, analytical skills and the ability to generalize previously learned information. Many animals have to learn, remember, generalize and solve various types of problems, such as negotiating their terrain, foreseeing the relation between cause and effect, escaping from dangers, and avoiding poisonous foods. Humans, too, have many different kinds of intelligence, such as the intelligence that makes someone a good mathematician, an effective CEO or a great basketball player.

Because learning and memory are two of the fundamental components of problem solving, it would not be totally surprising if enhancing learning and memory skills led to improved intelligence. But the various kinds of intelligence mean that the type and degree of enhancement must be highly dependent on the nature of the learning and memory skills involved in a particular task. Animals with an improved ability to recognize objects and

solve mazes in the laboratory, for instance, might have an easier time finding food and getting around from place to place in the wild. They might
also be more likely to escape from predators or even to learn to avoid traps.
But genetic engineering will never turn the mice into geniuses capable of
playing the piano.

Our finding that a minor genetic manipulation makes such a measurable
difference in a whole set of learning and memory tasks points to the possibility that NR2B may be a new drug target for treating various age-related
memory disorders. An immediate application could be to search for chemicals that would improve memory by boosting the activity or amount of
NR2B molecules in patients who have healthy bodies but whose brains have
begun to be ravaged by dementia during aging. Such drugs might improve
memory in mildly and modestly impaired patients with Alzheimer's disease
and in people with early forms of other dementias. The rationale would be
to boost the memory function of the remaining healthy neurons by modulating and enhancing the cells' NR2B activity. Of course, designing such
compounds will take at least a decade and will face many uncertainties. The
possible side effects of such drugs in humans, for example, would need to be
carefully evaluated, although the increased NR2B activity in the *Doogie*
mice did not appear to cause toxicity, seizures or strokes.

But if more NR2B in the brain is good for learning and memory, why
has nature arranged for the amount to taper off with age? Several schools of
thought weigh in on this question. One posits that the switch from NR2B to
NR2A prevents the brain's memory capacity from becoming overloaded.
Another, which I favor, suggests that the decrease is evolutionarily adaptive
for populations because it reduces the likelihood that older individuals—
who presumably have already reproduced—will compete successfully
against younger ones for resources such as food.

The idea that natural selection does not foster optimum learning and
memory ability in adult organisms certainly has profound implications. It
means that genetically modifying mental and cognitive attributes such as
learning and memory can open an entirely new way for the targeted genetic
evolution of biology, and perhaps civilization, with unprecedented speed.

The Basics: A Mouse Named *Doogie*

Joe Z. Tsien

How are *Doogie* mice different from other mice? They have been genetically engineered to make more than the usual amount of a key subunit of a protein called the N-methyl-D-aspartate (NMDA) receptor.

What does the NMDA receptor do? It helps to strengthen the connection between two neurons that happen to be active at the same time. Scientists theorize that such strengthening is the basis for learning and memory.

How smart are *Doogie* mice? They will never do differential equations or play the stock market, but they are better than normal mice at distinguishing between objects they have seen before and at recalling how to find a platform in a tank of murky water, for instance.

How does their genetic alteration make them smarter? The NMDA receptors of *Doogie* mice stay open nearly twice as long as those of normal mice. The extra time somehow helps them form a new memory more effectively.

Could the same technique be used to enhance people's ability to learn and remember? Theoretically, the possibility exists. But learning and memory in humans are much more complex than recognizing objects or remembering a water maze. Besides the scientific and technical barriers, the safety and ethical issues surrounding human genetic engineering would also need to be addressed. It is much more likely that pharmaceutical companies will first attempt to develop drugs that interact with the NMDA receptor to boost memory ability in people with memory deficits.

Testing *Doogie*: Putting the Smart Mouse through Its Paces

Joe Z. Tsien

In the initial tests of *Doogie* mice, we found that they were more likely than normal mice to recognize a familiar object over a novel one. But that test, which is called an object-recognition task, assesses only one type of memory.

To further evaluate whether *Doogie* mice have enhanced learning and memory abilities, we used a more complex laboratory test called the Morris water maze. In this test we put a mouse into a circular pool that was 1.2 meters in diameter and filled with murky water. We placed into the pool a nearly invisible, clear Plexiglas platform that was almost—but not quite—as tall as the water was deep, so that it was just hidden beneath the surface. We surrounded the pool with a black shower curtain that had certain landmarks on it, such as a red dot. Mice do not like to get wet, so in these tests they generally swim around until they find the platform, where they can pull themselves almost out of the water and rest.

We found that the *Doogie* mice located the submerged platform faster than normal mice, so we took the test a step further: we removed the platform to see if the animals would remember where the platform had been in relation to landmarks such as the red dot. When we put them back into the pool, *Doogie* mice spent more time than normal mice in the quarter of the pool where the platform had been, indicating that they remembered where it should be. What did they get as a reward? A toweling off and a stint under the heat lamp.

The Search for a Memory-Boosting Drug

Carol Ezzell

How close are researchers to devising a pill to help you remember where you put your car keys? The short answer is "not very." But that doesn't mean they aren't working on it—and hard. Less than eight months after Joe Z. Tsien of Princeton University (the author of the preceding articles) and his colleagues reported genetically engineering a smarter mouse, Tsien teamed up with venture capitalist Charles Hsu to form a company based on the discovery.

The newly incorporated firm is called Eureka Pharmaceuticals, and its home for the time being is Hsu's office at the Walden Group in San Francisco. The company's first order of business is to use gene technology called genomics to identify molecules that are potential targets for drugs to treat central nervous system disorders such as memory loss and dementia. "We believe the tools that Joe and his colleagues have developed can be translated pretty quickly into a basis for discovering therapies for human disease," Hsu says. Hsu is the CEO of Eureka; Tsien is the company's scientific adviser but will remain at Princeton.

Eureka's first target is the so-called NMDA receptor—which Tsien and his coworkers manipulated genetically to make their smart *Doogie* mice—although the company will also look for other targets. The receptor is essentially a pore that allows calcium to enter nerve cells, a prerequisite for strengthening the connection between two nerve cells. Such strengthening is thought to be the basis for learning and memory.

Over the past decade, several pharmaceutical companies have tested as possible stroke drugs various compounds that decrease the activity of the NMDA receptor. When the brain is starved of blood, such as happens when the blood clot of a stroke blocks an artery, nerve cells can release too much glutamate, a chemical the cells use to communicate. In a phenomenon called excitotoxicity, the excess glutamate binds to NMDA receptors on other nerve cells, allowing a tsunami of calcium to flood into the other cells. Together with the lack of oxygen, this causes the cells to die.

So far, however, the search for NMDA-receptor blockers that could serve as stroke drugs has been "incredibly disappointing," comments neuroscientist Robert C. Malenka of Stanford University. The problem, he explains, is finding a chemical that binds to precisely the right spot on the NMDA receptor and in just the right way, without causing other neurological effects. (After all, the illicit hallucinogenic drug phencyclidine—also known as PCP or "angel dust"—also binds to the receptor.)

The lack of success with NMDA receptor blockers against stroke— together with the possibility that agents that bind to the receptor might be toxic—has blunted some scientists' enthusiasm for developing drugs that might boost learning and memory by activating the receptor. "Nobody is seriously considering upregulating the activity of the NMDA receptor to boost memory, to my knowledge," Malenka says. "But maybe some clever person will come up with that magic drug that will tweak the receptor just so."

A more likely scenario—and one being pursued by Tsien—might be developing drugs that subtly modulate the activity of the NMDA receptor, without binding to it directly, according to Ira B. Black of the University of Medicine and Dentistry of New Jersey. Black studies a naturally occurring chemical called brain-derived neurotrophic factor (BDNF), which increases the likelihood that parts of the NMDA receptor will have a phosphate group tacked onto them. NMDA receptors with phosphate groups are more likely to be active than those without such groups.

Still, most neuroscientists concur that the search for a drug that enhances learning and memory without side effects will take time.

Combinatorial Chemistry and New Drugs

Matthew J. Plunkett and Jonathan A. Ellman

To fight disease, the immune system generates proteins known as antibodies that bind to invading organisms. The body can make about a trillion different antibodies, produced by shuffling and reshuffling their constituent parts. But the immune system is not equipped to craft a specialized antibody each time it is faced with a new pathogen. Instead the body selectively deploys only those existing antibodies that will work most effectively against a particular foe. The immune system does this, in effect, by mass screening of its antibody repertoire, identifying the ones that work best and making more of those. In the past few years, we and other chemists have begun to follow nature's example in order to develop new drugs. In a process called combinatorial chemistry, we generate a large number of related compounds and then screen the collection for the ones that could have medicinal value.

This approach differs from the most common way pharmaceutical makers discover new drugs. They typically begin by looking for signs of a desired activity in almost anything they can find, such as diverse collections of synthetic compounds or of chemicals derived from bacteria, plants or other natural sources. Once they identify a promising substance (known in the field as a lead compound), they laboriously make many one-at-a-time modifications to the structure, testing after each step to determine how the changes affected the compound's chemical and biological properties.

Often these procedures yield a compound having acceptable potency and safety. For every new drug that makes it to market in this way, however, researchers quite likely tinkered with and abandoned thousands of other compounds en route. The entire procedure is time-consuming and expensive: it takes many years and hundreds of millions of dollars to move from a lead compound in the laboratory to a bottle of medicine on the shelf of your local pharmacy.

The classical approach has been improved by screening tests that work more rapidly and reliably than in the past and by burgeoning knowledge about how various modifications are likely to change a molecule's biological activity. But as medical science has advanced, demand for drugs that can intervene in disease processes has escalated. To find those drugs, researchers need many more compounds to screen as well as a way to find lead compounds that require less modification.

Finding the Right Combination

Combinatorial chemistry responds to that need. It enables drug researchers to generate quickly as many as several million structurally related molecules. Moreover, these are not just any molecules, but ones that a chemist, knowing the attributes of the starting materials, expects will have a desired property. Screening of the resulting pool of compounds reveals the most potent varieties. Combinatorial chemistry can thus offer drug candidates that are ready for clinical testing faster and at a lower cost than ever before.

Chemists make combinatorial collections, or libraries, of screenable compounds in a rather simple way. We rely on standard chemical reactions to assemble selected sets of building blocks into a huge variety of larger structures. As a simplified example, consider four molecules: A1, A2, B1 and B2. The molecules A1 and A2 are structurally related and are thus said to belong to the same class of compounds; B1 and B2 belong to a second class. Suppose that these two classes of compounds can react to form molecules, some variant of which we suspect could produce a potent drug. The techniques of combinatorial chemistry allow us to construct easily all the possible combinations: A1-B1, A1-B2, A2-B1 and A2-B2.

Of course, in the real world, scientists typically work with much larger numbers of molecules. For instance, we might select 30 structurally related compounds that all share, say, an amine group (NH_2). Next, we might choose a second set of 30 compounds that all contain a carboxylic acid (CO_2H). Then, under appropriate conditions, we would mix and match every amine with every carboxylic acid to form new molecules called amides (CONH). The reaction of each of the 30 amines with each of the 30 carboxylic acids gives a total of 30 x 30, or 900, different combinations. If we were to add a third set of 30 building blocks, the total number of final structures would be 27,000 (30 x 30 x 30). And if we used more than 30 molecules in each set, the number of final combinations would rise rapidly.

Drugmakers have two basic combinatorial techniques at their disposal. The first, known as parallel synthesis, was invented in the mid-1980s by H. Mario Geysen, now at Glaxo Wellcome. He initially used parallel synthesis as a quick way to identify which small segment of any given large protein bound to an antibody. Geysen generated a variety of short protein fragments, or peptides, by combining multiple amino acids (the building blocks of peptides and proteins) in different permutations. By performing dozens or sometimes hundreds of reactions at the same time and then testing to see whether the resulting peptides would bind to the particular antibody of interest, he rapidly found the active peptides from a large universe of possible molecules.

In a parallel synthesis, all the products are assembled separately in their own reaction vessels. To carry out the procedure, chemists often use a so-called microtitre plate—a sheet of molded plastic that typically contains eight rows and 12 columns of tiny wells, each of which holds a few milliliters of the liquid in which the reactions will occur. The array of rows and columns enables workers to organize the building blocks they want to combine and provides a ready means to identify the compound in a particular well.

For instance, if researchers wanted to produce a series of amides by combining eight different amines and 12 carboxylic acids using the reactions we described earlier, they would place a solution containing the first amine in the wells across the first row, the second amine across the second row, and so on. They would then add each of the carboxylic acids to the wells sequentially, supplying a different version to each column. From only 20 different building blocks, investigators can obtain a library of 96 different compounds.

Chemists often start a combinatorial synthesis by attaching the first set of building blocks to inert, microscopic beads made of polystyrene (often referred to as solid support). After each reaction, researchers wash away any unreacted material, leaving behind only the desired products, which are still tethered to the beads. Although the chemical reactions required to link compounds to the beads and later to detach them introduce complications to the synthesis process, the ease of purification can outweigh these problems.

In many laboratories today, robots assist with the routine work of parallel synthesis, such as delivering small amounts of reactive molecules into the appropriate wells. In this way, the process becomes more accurate and less tedious. Scientists at Parke-Davis constructed the first automated method for parallel synthesis—a robotic device that can generate 40 compounds at a time. And investigators at Ontogen have developed a robot that can make up to 1,000 compounds a day. In general, the time needed to complete a parallel synthesis depends on how many compounds are being produced: when making thousands of compounds, doubling the number of products requires nearly twice as much time. Such practical considerations restrict parallel synthesis to producing libraries containing tens of thousands of compounds rather than many more.

Split and Mix

The second technique for generating a combinatorial library, known as a split-and-mix synthesis, was pioneered in the late 1980s by Árpád Furka, now at Advanced Chem Tech in Louisville, Kentucky. In contrast to parallel synthesis, in which each compound remains in its own container, a split-and-mix synthesis produces a mixture of related compounds in the same reaction vessel. This method substantially reduces the number of containers required and raises the number of compounds that can be made into the millions. The trade-off, however, is that keeping track of such large numbers of compounds and then testing them for biological activity can become quite complicated.

A simple example can explain this approach. Imagine that researchers have three sets of molecules (call them A, B and C), each set having three members (A1, A2, A3; B1, B2, B3; and so on). Inside one container, they attach the A1 molecules to polystyrene beads; in a second container A2 molecules, and in a third A3 molecules. Then the workers place all the bead-bound A molecules into one reaction vessel, mix them well and split them again into three equal portions, so that each vial holds a mixture of the three compounds. The researchers then add B1 molecules to the first container, B2 to the second, and B3 to the third. One more round of additions to introduce the C molecules produces a total of 27 different compounds.

To isolate the most potent of these structures, scientists first screen the final mixtures of compounds and determine the average activity of each batch. Then, using a variety of techniques, they can deduce which of the combinations in the most reactive batch has the desired biological activity.

A number of pharmaceutical companies have also automated the split-and-mix procedure. One of the earliest announcements came from a group at Chiron. Chemists there developed a robotic system that can make millions of compounds in a few weeks using this approach. The robot delivers chemicals and performs the mixing and partitioning of the solid support.

As we mentioned earlier, one of the problems with a split-and-mix synthesis is identifying the composition of a reactive compound within a large mixture. Kit Lam of the University of Arizona has developed a way to overcome this obstacle. He noted that at the end of a split-and-mix synthesis, all the molecules attached to a single bead are of the same structure. Scientists can pull out from the mixture the beads that bear biologically active molecules and then, using sensitive detection techniques, determine the molecular makeup of the compound attached. Unfortunately, this technique will work only for certain compounds, such as peptides or small segments of DNA.

Other investigators have developed methods to add to each bead a chemical label essentially listing the order in which specific building blocks have been added to the structure—in other words, the chemical equivalent of a UPC bar code. Reading the collection of these so-called tags on a particular bead gives a unique signature and hence the identity of the compound on that bead. Researchers at the biotechnology company Pharmacopeia, drawing on techniques introduced by W. Clark Still of Columbia University, have been very successful in applying powerful tagging techniques to their combinatorial libraries. Nevertheless, because of the difficulties of identifying

Split-and-Mix Synthesis

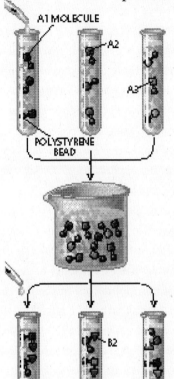

A1 MOLECULE

A2

A3

POLYSTYRENE
BEAD

STEP 1
Start with test tubes holding a solution containing inert polystyrene beads (gray circles). For simplicity, this example shows only three containers, but dozens might be used. Add the first set of molecules—the A class (squares)—to the test tubes, putting A1 molecules into the first container, A2 molecules into the second, and so on.

STEP 2
Mix the contents of all the test tubes.

B2

B3

B1 MOLECULE

STEP 3
Split the mixture into equivalent portions. Then add the second set of molecules—the B class (triangles)—placing B1 molecules into the first test tube, B2 into the second, et cetera. (Repeat steps 2 and 3 as many times as needed, depending on the number of sets of building blocks to be added.)

STEP 4
Separate the beads from any unreacted chemicals and detach the final structures. Researchers often screen the contents of each test tube to determine the mixture's average biological activity. Because each mixture shares the same final component, workers can determine which variant scores best—say, B2 might be most potent. They repeat the synthesis, adding only B2 to the A compounds to find which of the A-B2 combinations are the most biologically active.

compounds made in a split-and-mix synthesis, most pharmaceutical companies today continue to rely on parallel synthesis.

Drug Libraries

Both the parallel and the split-and-mix techniques of combinatorial chemistry began as ways to make peptides. Although these molecules are important in biological systems, peptides have limited utility as drugs because they degrade in the gut, they cannot be absorbed well through the stomach and they are rapidly cleared from the bloodstream. The pharmaceutical industry began to pursue combinatorial methods more aggressively after realizing that these techniques could also be applied to druglike compounds, such as the class of molecules known as benzodiazepines.

Benzodiazepines are one of the most widely prescribed classes of medicines. The best-known representative is diazepam, or Valium, but the class includes a number of other derivatives with important biological activity: anticonvulsant and antihypnotic agents, antagonists of platelet-activating factor (a substance important in blood clotting), inhibitors of the enzyme reverse transcriptase in HIV and inhibitors of Ras farnesyl transferase (an enzyme involved in cancer). Because of the broad activity of this class, benzodiazepines were the first compounds to be studied in a combinatorial synthesis seeking new drugs. In 1992 one of us (Ellman), working with Barry Bunin, also at the University of California at Berkeley, described a way to synthesize benzodiazepines on a solid support, making possible the synthesis of libraries containing thousands of benzodiazepine derivatives.

More recently, the two of us (Plunkett and Ellman) worked out a better approach for making benzodiazepines on a solid support; our new synthesis provides easy access to much larger numbers of compounds. The most challenging aspect of any combinatorial synthesis is determining the experimental conditions that will minimize side reactions giving rise to impurities. We spent more than a year fine-tuning the reaction conditions for our new benzodiazepine synthesis, but after determining the optimal procedure, we, along with Bunin, generated 11,200 compounds in two months using a parallel synthesis.

Promising Leads

From our benzodiazepine libraries, we have identified several compounds with promising biological activity. In a project with Victor Levin and Raymond Budde of the University of Texas M. D. Anderson Cancer Center in Houston, we have identified a benzodiazepine derivative that inhibits an enzyme implicated in colon cancer and osteoporosis. And in collaboration with Gary Glick and his colleagues at the University of Michigan, we discovered another benzodiazepine that inhibits the interaction of antibodies with single-strand DNA—a process that may be involved in systemic lupus erythematosus. These compounds are still in the very early stages of laboratory testing.

Once we and others demonstrated that combinatorial chemistry could be used to assemble druglike molecules, the pharmaceutical industry began pursuing more projects in this area. In the past five years, dozens of small companies devoted entirely to combinatorial chemistry have begun operation. Nearly all the major pharmaceutical companies now have their own combinatorial chemistry departments or have entered a partnership with a smaller, specialized company that does.

As might be expected, researchers have branched out beyond benzodiazepines, routinely applying combinatorial techniques to a wide array of starting materials. In general, chemists use combinatorial libraries of small organic molecules as sources of promising lead compounds or to optimize the activity of a known lead. When searching for a new lead structure, researchers often generate large libraries, with tens of thousands or even millions of final products. In contrast, a library designed to improve the potency and safety of an existing lead is typically much smaller, with only a few hundred compounds.

Several pharmaceutical companies are now conducting human clinical trials of drug candidates discovered through combinatorial chemistry. Because such programs are relatively new, none of these candidates has yet been studied long enough to receive approval from the U.S. Food and Drug Administration. But it is only a matter of time before a medicine developed with assistance from combinatorial methods reaches the market. Pfizer has one example in its pipeline. Using standard methods in 1993, the company discovered a lead compound that appeared to have potential for preventing atherosclerosis, or hardening of the arteries. In less than a year, using parallel synthesis, one laboratory at Pfizer generated in excess of 1,000 derivatives of the original structure, some of which were 100 times more potent than the lead compound. A drug derived from this series is now in human clinical trials. Notably, workers made more than 900 molecules before they noticed any improvement in biological activity. Few laboratories making standard one-at-a-time modifications to a lead compound could afford the time and money to generate nearly 1,000 derivatives that showed no advantage over the original substance.

Chemists at Eli Lilly also used parallel synthesis to develop a compound now in clinical trials for the treatment of migraine headaches. They had earlier found a lead substance that bound effectively to a desired drug target, or receptor. But the lead also had a high affinity for other, related receptors, behavior that could produce unwanted side effects. Researchers used parallel synthesis to make approximately 500 derivatives of that lead before arriving at the one currently being evaluated.

Researchers will inevitably find ways to generate combinatorial libraries even faster and at a lower cost. Already they are working out clever reaction methods that will enhance the final yield of products or replace the need for adding and later removing polystyrene beads. The future will also see changes in how information about the activity of tested compounds is gathered and analyzed in the pharmaceutical industry. For example, data on how thousands of compounds in one combinatorial library bind to a particular receptor can be used to predict the shape, size and electronic charge properties of that receptor, even if its exact structure is unknown. Such information can guide chemists in modifying existing leads or in choosing starting materials for constructing new combinatorial libraries.

Although the focus here has been the discovery of drugs, the power of combinatorial chemistry has begun to influence other fields as well, such as materials science. Peter G. Schultz and his colleagues at the University of California at Berkeley have used combinatorial methods to identify high-temperature superconductors. Other researchers have applied combinatorial techniques to liquid crystals for flat-panel displays and materials for constructing thin-film batteries. Scientists working on these projects hope to produce new materials quickly and cheaply. Clearly, the full potential of this powerful approach is only beginning to be realized. Combinatorial chemistry can appear somewhat random; combining various building blocks and hoping something useful comes out of the mix may seem to be the triumph of blind luck over knowledge and careful prediction. Yet this impression is far from the truth. A good library is the result of extensive development and planning. Chemists must decide what building blocks to combine and determine how to test the resulting structures for biological activity. Combinatorial chemistry allows researchers to gather, organize and analyze large amounts of data in a variety of new and exciting ways. The principle of selecting the most effective compounds from a collection of related ones—the guiding axiom of the immune system—is changing the way chemists discover new drugs. It is a pleasant irony that this lesson gleaned from our natural defenses can be helpful when those defenses fail.

Building the Better Bug

David A. O'Brochta and Peter W. Atkinson

An extraterrestrial visitor would surely acknowledge humanity to be the dominant species on the Earth. Should that visitor move past individual species and up the levels of taxonomic classification, however, the alien's field report might well give the class Insecta top billing. More than one million insect species have been identified, accounting for five-sixths of all species of animals. Each U.S. acre averages 400 pounds of insects, compared with only 14 pounds of people. Where humans and insects interact, vast economic interests hang in the balance. Even more profoundly, the clash of humans and insects that carry diseases is often a matter of life and death.

A few insect species, most notably those that feed on blood, are still responsible for spreading major human diseases, such as malaria, yellow fever, trypanosomiasis and dengue, as well as some conditions affecting livestock. Malaria alone accounts for between 300 million and 500 million clinical cases annually and some 1.5 million to 2.7 million deaths. About 200,000 people come down with yellow fever annually, and 30,000 die. Some 50 million people contract dengue every year; mortality can reach 15 percent without treatment. In many developing countries, nonfatal but debilitating conditions, such as dysentery, can be transmitted by insects, including the common housefly.

Public health efforts against insect-borne diseases have been limited to three basic strategies: rid the area entirely of the insect, use pesticides and physical barriers such as bed nets to keep at least some of the insects away, or develop a vaccine. The first undertaking has worked in some areas. Lowering the exposure to insects has had limited success. Vaccine development remains spotty; for example, the world still awaits an effective and affordable malaria vaccine.

An additional approach could cut this Gordian knot: simply make the insect unable to transmit disease. Insect bites themselves have little health consequence for most people; the pathogenic viruses, protozoa and filarial worms they transmit are the culprits. In the 1960s Chris F. Curtis of the London School of Hygiene and Tropical Medicine proposed that malaria could be stopped in its tracks if a way could be found genetically to convert its carrier, the *Anopheles* mosquito, to a form incapable of transmitting the *Plasmodium* protozoan actually responsible for the disease. Some mosquitoes are in fact naturally "refractory," or unable to transmit *Plasmodium*.

Curtis's proposal was impossible to implement for decades. But it soon may be realistic, thanks to modern genetic technologies. Genetic material from one species can be permanently integrated into the DNA of individuals from another species, conferring new traits. The resultant plant or animal that carries the new DNA is called transgenic.

Finding ways to engineer refractoriness into disease-carrying mosquitoes and other insect vectors now drives an extremely active area of research. The benefits of developing transgenic insects are not limited to medicine, however. The insertion of genes for useful products into the genomes of cows and goats has already created animals that produce pharmaceuticals in their milk. Applied to insects, transgenics could fundamentally change agriculture and the synthesis of some materials.

Jumping Genes Open Doors

The first glimmers of transgenic insect research date back to the 1960s, with the motivation being improved gene analysis rather than any direct applications outside the lab. Most of the early efforts to alter a genome consisted of injecting an insect egg with, or even simply bathing it in, DNA. Neither technique ever developed into a reliable method for producing transgenic insects.

All genetic research leapt forward in the early 1980s with the insect work of Gerald M. Rubin and Alan C. Spradling, both then at the Carnegie Institute of Washington. Rubin and Spradling were investigating fascinating genetic entities known as transposable elements. These strange strings of DNA have the ability to cut and paste themselves repeatedly into different chromosomes. Formally called transposons, their acrobatics also earned them the nickname "jumping genes." Geneticist Barbara McClintock discovered transposons during research in the 1940s on corn. The importance of her findings finally won her the Nobel Prize in 1983.

The particular transposon that the Carnegie researchers investigated came from the genome of that workhorse of the laboratory, the fruit fly *Drosophila melanogaster*. This species has little importance as an agricultural pest but has been fundamental in genetics research for the past 70 years. Acknowledging the propensity of the transposon to integrate itself into chromosomes, Rubin and Spradling had a simple and clever notion: Why not attach to it the gene they wanted the fly to have? They introduced an altered transposon into a cell, where it indeed pasted itself into the chromosome, creating transgenic *D. melanogaster*. The success and simplicity of their technique revolutionized the way researchers study the genetics and biology of that species.

The *Drosophila* transposon is known as the P element. It was discovered in the 1970s when geneticists noted a puzzling phenomenon. When males from certain populations mated with females of other populations, their progeny had numerous genetic aberrations, such as mutations, broken chromosomes and developmental abnormalities. Because the genetic entities responsible were discovered to come from only the paternal lineage, they were dubbed P factors. Eventually shown to be a transposon, the P factor of most interest to geneticists was renamed the P element and has proved priceless to *Drosophila* geneticists, allowing analysis of isolated genes and their effects.

Unfortunately, in 1986 a set of experiments by one of us (O'Brochta) and Alfred M. Handler of the U.S. Department of Agriculture in Gainesville, Florida, came to a frustrating conclusion: the P element is of little practical value beyond basic genetics research involving *D. melanogaster*. It will not readily insert itself into the chromosomes of other species. Ultimately, however, these experiments led to a new path, by shifting experimenters' attention to other transposable elements and strategies. Recent work has begun to uncover methods for creating transgenic insect species of greater importance.

The realization that the P element would not prove useful outside of

D. melanogaster sent biologists in search of more generally functional transposons. An obvious question became one of choice: Which transposable elements showed the most promise? Most researchers believed we should stick with a proved commodity and seek transposons that generally resembled the P element (called short inverted repeat-type transposable elements). Our lab developed techniques to determine quickly whether a particular transposon would successfully incorporate itself into the DNA of an insect species, which has helped speed the entire process of vector evaluation and development. Early efforts with transposons having structures similar to the P element have rewarded the choice to sail within sight of charted land.

In 1995 Charalambos Savakis and his colleagues at the Institute of Molecular Biology and Biotechnology on the Greek island of Crete succeeded in using a transposon called *Minos*, isolated from *D. hydei*. Using *Minos* to insert a novel gene into a Medfly, they created the first transgenic version of that animal. The transformation changed a fly with colorless eyes to one that expressed a gene for red-colored eyes (in a sense, effecting gene therapy). Subsequently, Handler and his coworkers successfully transformed Medflies with pigment-free eyes to ones having color, using a transposon called *piggy-Bac* that comes from the cabbage looper moth. Obviously, changing an insect's eye color is of little inherent interest; the importance of these groundbreaking successes is their illustration of the potential for creating transgenic insects that express truly valuable genes.

In 1998 investigators reported two discrete transformations of the *Aedes aegypti* mosquito, which transmits yellow fever and dengue. The successful manipulations of this inadvertently malevolent creature have led to greater optimism that geneticists will soon be able to convert it into a noncombatant in the disease wars. First, Anthony A. James and his colleagues at the University of California at Irvine altered *Ae. aegypti* via a housefly transposon called *Hermes*, which was originally isolated in our laboratory. (In contrast to the P element, *Hermes* appears to be an efficient vehicle for the creation of transgenic insects ranging from moths to mosquitoes. Work with it is helping to further the understanding of the biochemistry of transposon movement and regulation.) James and his research group then succeeded in incorporating into *Ae. aegypti* a transposon called *mariner*, isolated from the fruit fly species *D. mauritiana*. Again, the effect was simply to change eye color.

In research that dovetails with these two transgenic developments, Barry J. Beaty and his coworkers at Colorado State University demonstrated the feasibility of engineering refractoriness into *Ae. aegypti*. One way to get a host to express a gene it does not ordinarily have is to infect it with a virus carrying that novel DNA sequence. Beaty's team infected the mosquitoes with a nonpathogenic virus that included a gene that prevented the dengue virus from replicating in its host's salivary glands. The infection stops subsequent transmission.

Beaty's research shows that it is possible to create a refractory insect, and therefore no theoretical barriers exist to impede creation of such a creature via genetic insertion. An overarching problem remains, however. Merely waiting for a transgenic insect to pass on its new gene to huge numbers of descendants in a strictly Mendelian fashion—in which a parent possessing one copy of the gene contributes it to only half of his or her descendants—can be a lengthy process.

A more practical plan would spread the genetic change through large numbers of insects much more quickly. Fortunately, the basic components for quickly creating an entire insect population all carrying the key gene appear to be available. Once again, the transposon makes things possible.

Because of their propensity for jumping to new chromosomes, as well as making multiple copies of themselves, transposons burst free of the constraints of strict Mendelian heredity. A transformation event achieved by a scientist may have placed a single transposon, carrying a single key gene, into the genome of the target animal. But that transposon then may act as a free agent and spread itself throughout the genome. When that happens, more than half the offspring inherit the transgene. Within the relatively short time of a very few insect generations, most of the population expresses the trait.

This type of gene dispersal has been seen in nature. In fact, the P element itself appears to be a recent addition to *D. melanogaster*'s genome, most likely jumping over from *D. willistoni* no more than a century ago.

Other techniques may be effective at creating transgenic species. Human gene therapy is based in part on unique systems (derived from retroviruses but no longer infectious themselves) that are able to integrate genes into a new host's genome. (This is a true transgenic transformation, as opposed to Beaty's use of viruses to simply infect an individual organism that will henceforth produce the viral gene product.) These stripped-down viral vectors, known as pantropic pseudotyped retroviruses, can interact with virtually any cell from any organism. They have recently been used to create transgenic fish and clams and have successfully integrated genetic material into cultured mosquito cells. These viral vectors may yet prove their worth in efforts with whole insects.

Because the objective is to alter an insect's phenotype—the outward expression of its genetic makeup, or genotype—most workers have logically concentrated on ways to integrate foreign genetic material into the chromosomes of the host insect itself. But Frank F. Richards and his colleagues at Yale University are developing a clever approach that sabotages the cargo rather than the ship. Because virtually all insects harbor microbes, either as passive hitchhikers or active colonists, Richards is counting on transforming them rather than their hosts. Insects that carry engineered microbes are called paratransgenic, as the insect genome itself is untouched.

Richards and Charles B. Beard of the Centers for Disease Control and Prevention have produced paratransgenic, blood-sucking kissing bugs that can no longer transmit the trypanosome microbe responsible for Chagas' disease, which afflicts some 18 million South Americans with cardiovascular, gastrointestinal and neurological problems. The researchers isolated a bacterial symbiont from the kissing bugs and genetically modified it to secrete a protein that killed its trypanosome fellow travelers. The researchers placed the bacteria back in the insects, which then no longer served as hosts for the disease-causing protozoa. The genetically engineered symbionts successfully spread through a caged population of insects, converting them from vectors of a horrific disease to mere pests.

Agricultural Applications

Obviously, insects play crucial roles, both positive and negative, in agriculture. Most species are, in fact, beneficial or even essential. For example, honeybees are responsible for the pollination of $10 billion worth of produce in the U.S. Countless other species take part in nutrient recycling and help to maintain high soil quality. A small minority, however, compete with us for food. Since the development of agriculture, these pests have continually threatened our capacity to grow, harvest and store crops.

The concept of employing a genetic approach to deal with insect pests was first proposed in the 1940s by an American and a Russian. Edward F. Knipling of the USDA and Aleksandr S. Serebrovsky proposed similar schemes: inundate a pest population with sterile members of the same species. The majority of matings then become ineffectual. Serebrovsky's contribution was unknown in the West until 1968, when scientists, in particular Curtis, rediscovered it. Today this strategy is known as the sterile insect technique (SIT). Large numbers of insects can be sterilized, usually by ionizing radiation, and repeated infusions of such sterile organisms can wipe out a pest. The altered insect itself is, in effect, the weapon.

SIT is currently protecting parts of the agricultural economies of the U.S., Mexico, Guatemala, Chile, Argentina, Japan and Zanzibar, among other places. The strategy is ideal in locations as divergent as urban Los Angeles and tropical Zanzibar, where the aim is to eradicate a specific insect pest without harming the surrounding environment with chemical insecticides.

One of SIT's crowning achievements was the eradication of the New World screwworm, first from the southeastern U.S. and eventually from Mexico, by the 1970s. (The screwworm will receive little sympathy from most onlookers. It lays eggs in open wounds in livestock, which then hatch into larvae that consume the animals' flesh. In effect, the screwworm eats its victims alive.) The technique has recently been used successfully against the dreaded Medfly in the Los Angeles area, with greater economic benefit: a permanent Medfly infestation would cost California approximately $1.5 billion annually.

SIT depends on genetics, in the form of traditional breeding programs. For example, a strain of Medfly has been created that allows scientists to kill all the female embryos with a pulse of high temperature. This strain permits entomologists to mass-produce, sterilize and release only males. By

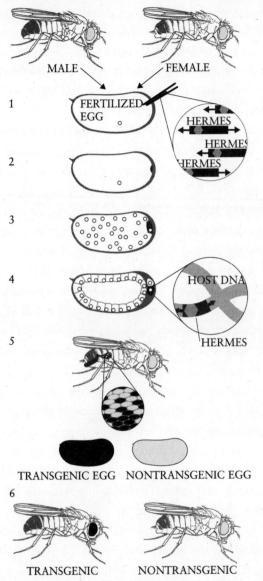

MALE   FEMALE

1   FERTILIZED EGG

HERMES

HERMES

HERMES

2

3

4   HOST DNA

HERMES

5

TRANSGENIC EGG   NONTRANSGENIC EGG

6

TRANSGENIC   NONTRANSGENIC

Making transgenic insects requires the insertion of a gene, carried by a transposable element such as *Heremes*, into a fertilized egg (*1*). The new genetic material is strategically placed at the polar plasm (*2*), that section of the egg destined to become the still nascent insect's own egg cells when it reaches maturity. After numerous divisions of the egg's nuclear material (*3*), most of it segregates to the periphery, where it will become the nuclei of the cells of the insect's body; two nuclei, however, will migrate to then become the insect's egg cells (*4*) when it reaches maturity (*5*). Should those cells have incorporated the transgene, progeny will be transgenic (*6*).

eliminating the females, the entire SIT program becomes more effective. Producing such strains, however, is difficult and time-consuming.

With transgenic technology, it should be possible to create insect strains in which only the females carry lethal genes. These deadly DNA segments would get expressed under specific conditions, such as the high-temperature pulse used in the traditional approach.

Transgenic technology also has the potential to reduce dependence on chemical pesticides. More than 900,000 American farms currently use insecticides, a reliance that can be counterproductive. Literally bathing insects in pesticides has actually driven the development of resistance to those killing agents. At least 183 species of insect and arachnid pests have developed resistance to one or more insecticides in the U.S. Furthermore, accumulation of chemical insecticides and their toxic breakdown products in our food, water, soil and textiles present a serious health issue. Biologists are therefore looking toward transgenics for the next wave of insect-control weapons.

One major problem with insecticides is that they kill many nontarget, beneficial species. They may wipe out useful predators and parasites, giving secondary pest species a chance to emerge. Transgenic technology may allow farmers to curtail pesticide use dramatically.

Currently a field can be sprayed with a wide array of different chemicals, each able to kill one or more harmful species. Unfortunately, the chemical controls against negative species often harm positive species as well. For example, mites attack California almond trees. One solution is to use an insecticide against those mites. Another is to release a species of predatory mite that kills the harmful mites. The trees, however, also fall victim to beetles and moths, so they are sprayed with other insecticides against those pests. And one of these insecticides kills the predatory mite that would free the tree from attack by the other mite species. A transgenic beneficial mite that could withstand the insecticides aimed at beetles and moths would allow farmers to spray fewer chemicals overall on their crops.

Of course, artificial selection for pesticide-resistant natural enemies is routine in labs, through conventional breeding practices. Brian A. Croft and his colleagues at the University of California at Riverside introduced insecticide-resistant predatory mites obtained from Washington State into orchards in southern California with good effect. Marjorie A. Hoy, now at the University of Florida, subsequently developed techniques that enabled the artificial selection of chemical resistance in natural predatory arthropods. Hoy and her coworkers developed an insecticide-resistant predatory mite that took part in the almond tree scenario just described.

This kind of effort, however, usually takes many insect generations and can result in a decrease in genetic diversity in these laboratory colonies when resistant progeny are bred with one another. The loss in genetic diversity can then be responsible for a lower general level of fitness outside the lab, in the field. This problem has limited the application of artificial selection for resistance in agricultural efforts to improve biological control. In contrast, transgenic insect technology can permit the rapid creation of resistant species. Natural enemies would become more practical and less expensive, making it more attractive to switch from chemical control.

Insecticide resistance is only the most obvious useful phenotype that could enhance the effectiveness of some beneficial arthropods. Other desirable qualities that could be engineered into insects with transgenic technology include pathogen resistance, general environmental hardiness, increased fecundity and improved host-seeking ability. One especially interesting application of transgenic engineering is the improvement of materials that insects supply to humans. Silk is a prime example. Breeding programs have created strains of silkworms that churn out more and better silk. Such programs, however, have limits: they cannot turn a silk purse into a steel one. But transgenic technology may. Some spider silks, for example, are stronger than Kevlar, a constituent of bulletproof vests and high-performance aircraft parts. Unfortunately, spider silk cannot be mass-produced as Kevlar is by humans or silk is by the larvae of silkworm. Transgenics offers the interesting possibility of introducing other species' silk genes into the genomes of silkworms. The creation of silkworms that spin bulletproof cocoons sounds fanciful, but such notions may be within the grasp of transgenic technology.

Ecological Considerations

The ability to engineer insects genetically confronts scientists and policymakers with questions about proper deployment, both in the laboratory and in the field. Guidelines already in place to govern the transfer of existing genetic expertise from the lab to the field can provide part of a regulatory framework. All technologies have risks associated with their use, and we need to develop an understanding of the specific risks inherent to insect transgenics and how to minimize them. For example, unintentional movement of transgenes between insect species is of some concern and is being investigated. In some cases, the risk should be minimal: when only sterile members of the transgenic species are released, the inserted genetic sequences should remain sequestered from the overall gene pool.

As the term "risk-benefit" implies, the good that could come from transgenics should be taken into account in decisions about development and deployment. Government agencies, such as the USDA's Animal and Plant Health Inspection Service, have established mechanisms to ensure that field releases of transgenic insects are done strictly with regard to genetic an ecological consequences. (Readers can review this implementation process, as well as the present application for field trails of transgenic arthropods, by visiting the USDA Web site at www.aphis.usda.gov/bbep.)

The genetic transformation of a given organism has, in every case, revolutionized the study of its biology. Furthermore, research in one species may inform work in others. Mendel's humble pea plants broke the ground, and studies of *Drosophila* laid the foundation for most of modern genetic research. Investigations with *Escherichia coli* led to pioneering research into regulatory mechanisms that hastened the development of genetic engineering technology in bacteria, plants and animals. Transgenic mammals already produce a variety of medical products. Applied to insects, transgenic technology can offer biologists new ways to investigate, control and exploit these creatures that hold sway over such a substantial proportion of human affairs.

Suggested Reading

Broach, James R., and Jeremy Thorner. "High-Throughput Screening for Drug Discovery," *Nature* 384, supplement, no. 6604 (November 7, 1996): 14–16.

Collins, Frank H., and Anthony A. James. "Genetic Modifications of Mosquitoes," *Science & Medicine* (November/December 1996): 52–61.

"Combinatorial Chemistry," special report in *Chemical & Engineering News* 74, no. 7 (February 12, 1996): 28–73.

Curtis, C. F., and P. M. Graves. "Methods for Replacement of Malaria Vector Populations," *Journal of Tropical Medicine and Hygiene* 91, no. 2 (April 1988): 43–48.

Czanik, Anthony W., and Jonathan A. Ellman, ed. "Combinatorial Chemistry," special issue of *Accounts of Chemical Research* 29, no. 3 (March 1996).

Hebb, Donald O. *The Organization of Behavior: A Neuropsychological Theory*. New York: John Wiley, 1949.

Kidwell, Margaret G., and Alice R. Wattam. "An Important Step Forward in the Genetic Manipulation of Mosquito Vectors in Human Disease," *Proceedings of the National Academy of Sciences USA* 95, no. 7 (March 31, 1998): 3349–3350.

Malenka, Robert C., and Roger A. Nicoll. "Long-Term Potentiation—A Decade of Progress?" *Science* 285, no. 5435 (September 17, 1999): 1870–1874.

Squire, Larry R. *Memory and Brain*. London: Oxford University Press, 1987.

Symondson, W. O. C., and J. E. Liddell. *The Ecology of Agricultural Pests: Biochemical Approaches*. New York: Chapman and Hall, 1996.

Thompson, Lorin A., and Jonathan A. Ellman. "Synthesis and Applications of Small Molecule Libraries," *Chemical Reviews* 96, no. 1 (January 1996): 555–600.

Tsien, Joe Z. "Enhancing the Link between Hebb's Coincidence Detection and Memory Formation," *Current Opinion in Neurobiology* 10, no. 2 (April 2000).

8 Taking Humankind to the Extreme

Head Transplants

Robert J. White

Livers, lungs, hearts, kidneys and, most recently, hands. With such rapid advances in the field of human transplantation, researchers such as myself are now beginning to consider what some have previously deemed unthinkable: transplanting a human brain.

I predict that what has always been the stuff of science fiction—the Frankenstein legend, in which an entire human being is constructed by sewing various body parts together—will become a clinical reality early in the 21st century. Our modern-day version of the tale will include the transplantation of the human brain with all its complexity preserved. But the brain can't function properly without the plumbing of the body and the wiring of the head. So brain transplantation, at least initially, will really be head transplantation—or body transplantation, depending on your perspective.

The concept of head transplantation has always held a certain fascination for experimental surgeons. As early as 1908, American physiologist and pharmacologist Charles C. Guthrie grafted the head of a small mixed-breed dog onto the neck of a larger one whose own head remained intact. Similarly, in the 1950s Russian scientist Vladimir P. Demikhov transplanted the upper body of a mixed-breed puppy—including the forelimbs—to the neck of a much larger dog by connecting the pup to the other dog's neck blood vessels. At least one of Demikhov's famous "two-headed dogs" reportedly survived as long as 29 days after the surgery.

It was not until 1970, however, that a mammalian head was successfully transplanted onto a mammalian body that had already had its own head removed. This was first accomplished by my colleagues and me in a nonhuman primate—a rhesus monkey. When the monkey awakened from anesthesia, it regained full consciousness and complete cranial nerve function, as measured by its wakefulness, aggressiveness, and ability to eat and to follow people moving around the room with its eyes. Such monkeys lived for as long as eight days. With the significant improvements in surgical techniques and postoperative management since then, it is now possible to consider adapting the head-transplant technique to humans.

A surgical protocol for head transplantation in humans would require very little alteration from that used in monkeys, although it would need to be scaled up because of the difference in body size between the two species. In fact, the procedure would be easier to perform in humans than in monkeys, because the blood vessels and other tissues of a human are larger than those of a monkey, and surgeons have much more experience operating on the human anatomy.

Maintaining an adequate, uninterrupted flow of blood to the brain would be absolutely essential during all stages of a human head-transplant operation because the brain, unlike other solid organs, cannot survive being separated from its blood supply (at least at normal body temperature). Surgeons would monitor the brain's activity—an indirect way to assess blood flow—during the procedure using electroencephalograph electrodes placed on the scalp. Each patient's head would also be placed in a circular clamp to allow it to be stabilized and moved safely.

Heads Off to You

The procedure would be conducted in a specially designed operating suite that would be large enough to accommodate equipment for two operations conducted simultaneously by two separate surgical terms. Once the two patients were anesthetized, the two teams, working in concert, would make deep incisions around each patient's neck, carefully separating all the tissues and muscles to expose the carotid arteries, jugular veins and spine. The surgeons would then place catheters coated with heparin, a drug that prevents blood clotting, into each of the blood vessels to ensure that the brain received sufficient blood flow and, therefore, oxygen. After removing bone from the spine of each patient's neck, they would cut open the protective membranes surrounding the spinal cord, exposing it. Following separation of the spine and cord, the head of one patient would be removed and transferred to the tubes that would connect it to the circulation of the second patient's body, which would have had its own head removed.

Once this critical maneuver was completed, the blood vessel tubes would be removed one by one, and the surgeons would sew the arteries and veins of the transplanted head together with those of the new body. The spinal columns would then be fastened together with metal plates, and the muscles and skin would be sewn together layer by layer.

My colleagues and I have already taken the first steps toward human head transplantation. We have developed pumps and devices to lower to 10 degrees Celsius (50 degrees Fahrenheit) the blood circulating to the head that is being prepared for transplantation. Such cooling slows the metabolism of the brain so that its blood supply can be cut off for up to an hour during surgery. The greatest hurdle remaining is how to prevent the body from rejecting the new head, and vice versa. It is unclear at this point whether the drugs now used to prevent rejection following transplantation of organs such as livers and kidneys will work for an entire body.

Longer Life for the Paralyzed?

Who might benefit from a head transplant? The first candidates for the procedure will probably be people who have been paralyzed from the neck down because of an accident. For reasons that are still unclear, such individuals often die prematurely of multiple-organ failure. Although transferring a paralyzed person's head to another body would not—at least at this point in the development of the technology—allow him to move or walk again, it could prolong his life. And many hope that in the twenty-first century, physicians will find a way to heal severed spinal cords, so those who have their heads transplanted onto a new body might someday receive sensory information from and gain motor control over it.

Where will bodies for head transplantation come from? The recipient body would be someone who has been declared brain dead. Such individuals already serve as multiple-organ donors, so there should be no strikingly new bioethics considerations for head transplantation.

But how well will we as a society accept the concept that human brain transplantation involves transplanting the mind and spirit? Are we willing to acknowledge that the human brain is the physical repository of the soul, something this operation implies? These are the questions facing us as we go in reality where Mary Wollstonecraft Shelley went only in fiction.

The Coming Merging of Mind and Machine

Ray Kurzwell

Sometime early in this century, the intelligence of machines will exceed that of humans. Within several decades, machines will exhibit the full range of human intellect, emotions and skills, ranging from musical and other creative aptitudes to physical movement. They will claim to have feelings and, unlike today's virtual personalities, will be very convincing when they tell us so. By 2019 a $1,000 computer will at least match the processing power of the human brain. By 2029 the software for intelligence will have been largely mastered, and the average personal computer will be equivalent to 1,000 brains.

Once computers achieve a level of intelligence comparable to that of humans, they will necessarily soar past it. For example, if I learn French, I can't readily download that learning to you. The reason is that for us, learning involves successions of stunningly complex patterns of interconnections among brain cells (neurons) and among the concentrations of biochemicals, known as neurotransmitters, that enable impulses to travel from neuron to neuron. We have no way of quickly downloading these patterns. But quick downloading will allow our nonbiological creations to share immediately what they learn with billions of other machines. Ultimately, nonbiological entities will master not only the sum total of their own knowledge but all of ours as well.

As this happens, there will no longer be a clear distinction between human and machine. We are already putting computers—neural implants—directly into people's brains to counteract Parkinson's disease and tremors from multiple sclerosis. We have cochlear implants that restore hearing. A retinal implant is being developed in the U.S. that is intended to provide at least some visual perception for some blind individuals, basically by replacing certain visual-processing circuits of the brain. Recently scientists from

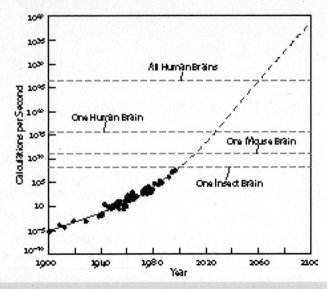

The accelerating rate of progress in computing is demonstrated by this graph, which shows the amount of computing speed that $1,000 (in constant dollars) would buy, plotted as a function of time. Computer power per unit cost is now doubling every year.

Emory University implanted a chip in the brain of a paralyzed stroke victim that allows him to use his brainpower to move a cursor across a computer screen.

In the 2020s neural implants will improve our sensory experiences, memory and thinking. By 2030, instead of just phoning a friend, you will be able to meet in, say, a virtual Mozambican game preserve that will seem compellingly real. You will be able to have any type of experience— business, social, sexual—with anyone, real or simulated, regardless of physical proximity.

How Life and Technology Evolve

To gain insight into the kinds of forecasts I have just made, it is important to recognize that technology is advancing exponentially. An exponential process starts slowly, but eventually its pace increases extremely rapidly. (A fuller documentation of my argument is contained in my book *The Age of Spiritual Machines*.)

The evolution of biological life and the evolution of technology have both followed the same pattern: they take a long time to get going, but advances build on one another and progress erupts at an increasingly furious pace. We are entering that explosive part of the technological evolution curve right now.

Consider: It took billions of years for Earth to form. It took two billion more for life to begin and almost as long for molecules to organize into the first multicellular plants and animals about 700 million years ago. The pace of evolution quickened as mammals inherited Earth some 65 million years ago. With the emergence of primates, evolutionary progress was measured in mere millions of years, leading to *Homo sapiens* perhaps 500,000 years ago.

The evolution of technology has been a contribution of the evolutionary process that gave rise to us—the technology-creating species—in the first place. It took tens of thousands of years for our ancestors to figure out that sharpening both sides of a stone created useful tools. Then, early in the last millennium, the time required for a major paradigm shift in technology had shrunk to hundreds of years.

The pace continued to accelerate during the 19th century, during which technological progress was equal to that of the 10 centuries that came before it. Advancement in the first two decades of the 20th century matched that of the entire 19th century. Today significant technological transformations take just a few years; for example, the World Wide Web, already a ubiquitous form of communication and commerce, did not exist just a decade ago.

Computing technology is experiencing the same exponential growth. Over the past several decades, a key factor in this expansion has been described by Moore's Law. Gordon Moore, a cofounder of Intel, noted in the mid-1960s that technologists had been doubling the density of transistors on integrated circuits every 12 months. This meant computers were periodically doubling both in capacity and in speed per unit cost. In the mid-1970s Moore revised his observation of the doubling time to a more accurate estimate of about 24 months, and that trend persisted through the 1990s.

After decades of devoted service, Moore's Law will have run its course around 2019. By that time, transistor features will be just a few atoms in width. But new computer architectures will continue the exponential growth of computing. For example, computing cubes are already being designed that will provide thousands of layers of circuits, not just one as in today's computer chips. Other technologies that promise orders-of-magnitude increases in computing density include nanotube circuits built from carbon atoms, optical computing, crystalline computing and molecular computing.

We can readily see the march of computing by plotting the speed (in instructions per second) per $1,000 (in constant dollars) of 49 famous calculating machines spanning the 20th century. The graph on page 208 is a study in exponential growth: computer speed per unit cost doubled every three years between 1910 and 1950 and every two years between 1950 and 1966 and is now doubling every year. It took 90 years to achieve the first $1,000 computer capable of executing one million instructions per second (MIPS). Now we add an additional MIPS to a $1,000 computer every day.

Why Returns Accelerate

Why do we see exponential progress occurring in biological life, technology and computing? It is the result of a fundamental attribute of any evolutionary process, a phenomenon I call the Law of Accelerating Returns. As order exponentially increases (which reflects the essence of evolution), the time between salient events grows shorter. Advancement speeds up. The returns—the valuable products of the process—accelerate at a nonlinear rate. The escalating growth in the price performance of computing is one important example of such accelerating returns.

A frequent criticism of predictions is that they rely on an unjustified extrapolation of current trends, without considering the forces that may alter those trends. But an evolutionary process accelerates because it builds on past achievements, including improvements in its own means for further evolution. The resources it needs to continue exponential growth are its own increasing order and the chaos in the environment in which the evolutionary process takes place, which provides the options for further diversity. These two resources are essentially without limit.

The Law of Accelerating Returns shows that by 2019 a $1,000 personal computer will have the processing power of the human brain—20 million billion calculations per seconds. Neuroscientists came up with this figure by taking an estimation of the number of neurons in the brain, 100 billion, and multiplying it by 1,000 connections per neuron and 200 calculations per second per connection. By 2055, $1,000 worth of computing will equal the processing power of all human brains on Earth (of course, I may be off by a year or two).

Programming Intelligence

That's the prediction for processing power, which is a necessary but not sufficient condition for achieving human-level intelligence in machines. Of greater importance is the software of intelligence.

One approach to creating this software is to painstakingly program the rules of complex process. We are getting good at this task in certain cases; the CYC (as in "encyclopedia") system designed by Douglas B. Lenat of Cycorp has more than one million rules that describe the intricacies of human common sense, and it is being applied to Internet search engines so that they return smarter answers to our queries.

Another approach is "complexity theory" (also known as chaos theory) computing, in which self-organizing algorithms gradually learn patterns of information in a manner analogous to human learning. One such method, neural nets, is based on simplified mathematical models of mammalian neurons. Another method, called genetic (or evolutionary) algorithms, is based on allowing intelligent solutions to develop gradually in a simulated process of evolution.

Ultimately, however, we will learn to program intelligence by copying the best intelligent entity we can get our hands on: the human brain itself. We will reverse-engineer the human brain, and fortunately for us it's not even copyrighted!

The most immediate way to reach this goal is by destructive scanning: take a brain frozen just before it was about to expire and examine one very thin slice at a time to reveal every neuron, interneuronal connection and concentration of neurotransmitters across each gap between neurons (these gaps are called synapses). One condemned killer has already allowed his brain and body to be scanned, and all 15 billion bytes of him can be accessed on the National Library of Medicine's Web site (www.nlm.nih.gov/research/visible/visible_gallery.html). The resolution of these scans is not nearly high enough for our purposes, but the data at least enable us to start thinking about these issues.

We also have noninvasive scanning techniques, including high-resolution magnetic resonance imaging (MRI) and others. Their increasing resolution and speed will eventually enable us to resolve the connections between neurons. The rapid improvement is again a result of the Law of Accelerating Returns, because massive computation is the main element in higher-resolution imaging.

Another approach would be to send microscopic robots (or "nanobots") into the bloodstream and program them to explore every capillary, monitoring the brain's connections and neurotransmitter concentrations.

Fantastic Voyage

Although sophisticated robots that small are still several decades away at least, their utility for probing the innermost recesses of our bodies would be far-reaching. They would communicate wirelessly with one another and report their findings to other computers. The result would be a noninvasive scan of the brain taken from within.

Most of the technologies required for this scenario already exist, though not in the microscopic size required. Miniaturizing them to the tiny sizes needed, however, would reflect the essence of the Law of Accelerating Returns. For example, the translators on an integrated circuit have been shrinking by a factor of approximately 5.6 in each linear dimension every 10 years.

The capabilities of these embedded nanobots would not be limited to passive roles such as monitoring. Eventually they could be built to communicate directly with the neuronal circuits in our brains, enhancing or extending our mental capabilities. We already have electronic devices that can communicate with neurons by detecting their activity and either triggering nearby neurons to fire or suppressing them from firing. The embedded nanobots will be capable of reprogramming neural connections to provide virtual-reality experiences and to enhance our pattern recognition and other cognitive faculties.

To decode and understand the brain's information-processing methods (which, incidentally, combine both digital and analog methods), it is not necessary to see every connection, because there is a great deal of redundancy within each region. We are already applying insights from early stages of this reverse-engineering process. For example, in speech recognition, we have already decoded and copied the brain's early stages of sound processing.

Perhaps more interesting than this scanning-the-brain-to-understand-it approach would be scanning the brain for the purpose of downloading it. We would map the locations, interconnections and contents of all the neurons, synapses and neurotransmitter concentrations. The entire organization, including the brain's memory, would then be re-created on a digital-analog computer.

To do this, we would need to understand local brain processes, and progress is already under way. Theodore W. Berger and his coworkers at the University of Southern California have built integrated circuits that precisely match the processing characteristics of substantial clusters of neurons. Carver A. Mead and his colleagues at the California Institute of Technology have built a variety of integrated circuits that emulate the digital-analog characteristics of mammalian neural circuits.

Developing complete maps of the human brain is not as daunting as it may sound. The Human Genome Project seemed impractical when it was first proposed. At the rate at which it was possible to scan genetic codes 12 years ago, it would have taken thousands of years to complete the genome. But in accordance with the Law of Accelerating Returns, the ability to sequence DNA has been accelerating. The latest estimates are that the entire human genome will be completed in just a few years.

By the third decade of the 21st century, we will be in a position to create complete, detailed maps of the computationally relevant features of the human brain and to re-create these designs in advanced neural computers. We will provide a variety of bodies for our machines, too, from virtual bodies in virtual reality to bodies comprising swarms of nanobots. In fact, humanoid robots that ambulate and have lifelike facial expressions are already being developed at several laboratories in Tokyo.

Will It Be Conscious?

Such possibilities prompt a host of intriguing issues and questions. Suppose we scan someone's brain and reinstate the resulting "mind file" into a suitable computing medium. Will the entity that emerges from such an operation be conscious? This being would appear to others to have very much the same personality, history and memory. For some, that is enough to define consciousness. For others, such as physicist and author James Trefil, no logical reconstruction can attain human consciousness, although Trefil concedes that computers may become conscious in some new way.

At what point do we consider an entity to be conscious, to be self-aware, to have free will? How do we distinguish a process that is conscious from one that just acts *as if* it is conscious? If the entity is very convincing when it says, "I'm lonely, please keep me company," does that settle the issue?

If you ask the "person" in the machine, it will strenuously claim to be the original person. If we scan, let's say, me and reinstate that information into a neural computer, the person who emerges will think he is (and has been) me (or at least he will act that way). He will say, "I grew up in Queens, New York, went to college at M.I.T., stayed in the Boston area, walked into a scanner there and woke up in the machine here. Hey, this technology really works."

But wait, is this really me? For one thing, old Ray (that's me) still exists in my carbon-cell-based brain.

Will the new entity be capable of spiritual experiences? Because its brain processes are effectively identical, its behavior will be comparable to that of the person it is based on. So it will certainly claim to have the full range of emotional and spiritual experiences that a person claims to have.

No objective test can absolutely determine consciousness. We cannot objectively measure subjective experience (this has to do with the very nature of the concepts "objective" and "subjective"). We can measure only correlates of it, such as behavior. The new entities will appear to be conscious, and whether or not they actually are will not affect their behavior. Just as we debate today the consciousness of nonhuman entities such as animals, we will surely debate the potential consciousness of nonbiological intelligent entities. From a practical perspective, we will accept their claims. They'll get mad if we don't.

Before the next century is over, the Law of Accelerating Returns tells us, Earth's technology-creating species—us—will merge with our own technology. And when that happens, we might ask: what is the difference between a human brain enhanced a millionfold by neural implants and a nonbiological intelligence based on the reverse-engineering of the human brain that is subsequently enhanced and expanded?

The engine of evolution used its innovation from one period (humans) to create the next (intelligent machines). The subsequent milestone will be for the machines to create their own next generation without human intervention.

An evolutionary process accelerates because it builds on its own means for further evolution. Humans have beaten evolution. We are creating intelligent entities in considerably less time than it took the evolutionary process that created us. Human intelligence—a product of evolution—has transcended it. So, too, the intelligence that we are now creating in computers will soon exceed the intelligence of its creators.

Nosing Out a Mate

Meredith F. Small

It's Saturday night in the year 2030 and time for your night on the town. It doesn't matter much what you wear; just be sure to dab on a little of that stuff you bought from the local pheromone shop. You might reach for a vial of your own essence that's been specially concentrated to make the most of your own attractive powers—or maybe you favor a synthesized version of the movie star of the moment's *je ne sais quoi*. Perhaps you go for a tube of the chemistry of some unknown person who just happens to be better-looking, more confident or blessed with superior genes to yours. Regard-less, it's off to the neighborhood Fern-and-Sniff bar, and good luck!

Recent research suggests that humans, like many other organisms, can be sensitive to pheromones, which are thought to be odorless chemicals secreted from the body and picked up by a special organ in the nose. In the animal kingdom and among insects, pheromones convey information to other members of the species about an individual's gender, reproductive sta-tus and rank on the social ladder. Contrary to popular misunderstanding, pheromones are not strictly sex attractants, but they do play a role in the mating rituals of everything from moths to mice.

Do humans have pheromones? Right now the jury is still out. But scien-tists know that something—perhaps a pheromone—in the underarm sweat of some women can alter the menstrual cycles of other women who come in close contact with them. Some investigators even have early indications that such airborne chemicals might unconsciously influence who we choose as mates.

More than a few researchers predict that science will isolate an incontro-vertible human pheromone early in the next century—in fact, some contend they already have. How will that change tomorrow's battle of the sexes? Will a chemical advantage in the mating dance be as close as the corner shop?

Other animals have been using their noses to find mates for a long time. Far up in the nasal passages of all mammals—including humans—are receptors that react to odors and pass on the signals we register as smells to the neocortex, the "gray matter" of the brain. But many claim we also possess another nasal sense called the vomeronasal organ (VNO), a pair of tiny sacs that lies closer to the nostrils. Receptor cells in the VNO supposedly pick up pheromones—which generally cannot be detected as a smell—and transmit the information to the hypothalamus and the limbic system, more primitive parts of the brain. These portions control the urges for such things as food and sex.

In the 1970s scientists showed that smell, whether of odors or pheromones, has a powerful role in mate choice—at least in rodents. Rodent urine, it seems, differs in odor and pheromone content according to what type of major histocompatibility complex (MHC) genes the animal has. MHC genes contain instructions for making the proteins that help an organism tell what belongs in its body and what is a potentially dangerous foreigner. Every rat or mouse (and, maybe, human) has its own chemical signature dictated by the MHC genes.

Interestingly, given their druthers, rodents sniff and then select mates with MHC genes that are quite different from their own. Such choices make good evolutionary sense: by choosing a mate whose MHC genes differ most markedly, rats and mice also have a good chance of mating with a partner whose other genes vary from theirs. Going for such variety could translate into offspring that are more equipped to fight off a range of infectious diseases.

If—and how—humans emit and track pheromones is not exactly clear. We belong to the order Primates, whose members are known for their well-developed senses of vision and touch. The trade-off for such visual and tactile acuity has supposedly been a less than keen sense of smell—a drawback that also seems to have blunted our pheromone-sensing abilities. In fact, most medical textbooks dismiss the human VNO as a vestigial structure that appears in the fetus and then nearly disappears later in development. But in the 1980s anatomists found evidence that the VNO exists in most adult people, even though it might not operate as well as it does in other mammals.

Something in the Air

More recent behavioral studies, however, suggest that the human VNO functions just fine. In 1998 Martha K. McClintock and Kathleen Stern, scientists at the University of Chicago, showed that substances isolated from the sweat of women at various phases of the menstrual cycle can modulate the timing of ovulation in women with whom they associate.

McClintock first documented in the 1970s that the menstrual cycles of women who spend a lot of time together eventually synchronize, suggesting that something that can waft from one woman to another must be at work. Researchers then used cotton pads to see if they could swab the substance from female armpits. They first removed any odoriferous compounds from the pads and then wiped the remaining tasteless, odorless liquid onto the upper lips of other women. After a few months, they observed, the periods of the women who agreed to have the potential pheromone dabbed under their nose were in sync.

The same research protocol was used to show that men can influence the female cycle as well. A group of men offered up their armpit sweat, which was deodorized and then wiped on the upper lips of women with irregular menstrual cycles. Repeated exposure to the male secretions caused the women to cycle regularly, presumably by making them ovulate in a timely manner. Living with a man is presumed to have the same effect.

Although the studies substantiate the power of putative pheromones on female physiology, they say nothing about the role of the compounds in choosing a mate. For that, researchers have been turning to what might seem the least likely subjects—members of a closed religious sect in the Midwest that proscribes extramarital sex.

Carole Ober of the University of Chicago turned to the Hutterites, who routinely marry within their group, because she was intrigued by the rodent studies done in the 1970s. She wondered if humans, too, might subconsciously tend to mate with someone who has a differing MHC gene profile, which in humans is called the human leukocyte antigen (HLA) system.

Ober found that even though Hutterites as a group have very similar HLA profiles because of their history of intermarriage, the HLA genes of the wives she and her colleagues studied were quite different from those of their husbands. This suggested to Ober that the Hutterites were unwittingly optimizing the genetic diversity of their children by marrying partners whose genes were least like their own.

But how did the Hutterite couples choose partners with such different genes? Ober thinks the answer may lie in pheromones. Other studies have demonstrated that humans can smell the difference between mice with different MHC genes, she notes. So maybe the elusive chemistry that brings certain people together really is a pheromone. "I think this is likely," Ober says. "It would be odd if we could discriminate among the MHC types of another species but not among our own kind."

In 1995 evolutionary biologist Claus Wedekind of the University of Bern tested the possibility. He determined the HLA types of 49 women and 44 men (who were unknown to one another) and then asked each man to wear a cotton T-shirt for two consecutive nights. Next he asked the women, most of whom were ovulating and presumably at their most perceptive for choosing a mate, to sniff the shirts and record their reactions.

Wedekind reported that the women tended to prefer shirts worn by men with HLA types dissimilar from theirs. What is more, the shirts reminded them of current or former boyfriends. The women found T-shirts that had been worn by men with HLA types similar to their own unattractive and commented that they smelled like their fathers or brothers.

That Come-Hither Smell?

Did Wedekind's T-shirts contain human pheromones that either attracted or repelled the women? And, as many will want to know, how soon can the substance be bottled and sold?

Whether anyone has identified and purified an actual human pheromone is the subject of heated debate among people who know about the nose. For his part, David Berliner of Pherin in Menlo Park, California, claims that his company has produced two colognes based on human pheromones: one for men and one for women. But even Berliner isn't touting the potions as sex attractants. In fact, the men's cologne contains what he claims are male pheromones and the female scent, female essence. He says the colognes are intended to make the wearer relaxed and self-confident, which will draw in members of the opposite sex—a theory, by the way, that still hasn't been clinically tested. In any case, the current design might work just fine for gays and lesbians, and enterprising heterosexuals might try simply switching bottles.

Berliner isn't the only one purporting to sell human pheromones. Winifred B. Cutler of the Athena Institute for Women's Wellness Research has branched into commercial products as well. Her Chester Springs, Pennsylvania–based company, which also conducts research, advertises vials of odorless synthetic human pheromones as additives to one's favorite scent. These scents are intended to "increase the romantic attention from the opposite sex," according to the ads.

Cutler asserts she has backed up this claim with a double-blind study of her men's solution. Men who used the compound in their aftershave lotion for six weeks reported that they increased the number of times they slept next to a woman and also said they had more sexual intercourse, she says. Because the men didn't masturbate more, she and her colleagues contend that the increase was not caused by heightened sex drive but by increased sex appeal.

What do other scientists think about all this? Even if Berliner and Cutler have isolated human pheromones—a point that is hotly contested—the stuff still might not matter when it comes to picking whom to bed down with. "If we do find an effect of pheromones on mate choice," McClintock comments, "I believe the role will be modulatory, that is, in concert with existing mechanisms that are already rich, complicated and dependent on context."

Does this mean that if we can bottle our chemistry and dole it out in the future, the additive will change the way we fall in love? No, flowers and candlelight and sweet nothings in the ear will still be important, according to most accounts.

We are a behaviorally fickle species. When it comes to finding a mate, we are swayed by culture, pushed by family and locked into traditions. In many places across the globe, people even have their mates chosen for them, pheromones be damned. We also sidestep biology by washing off our body odors and any pheromones or diluting them with soap and perfume.

Perhaps in the future we will be able to better control the messy process of the mating dance with a touch of something that makes us especially appealing to others. That way we could concentrate on projecting the good points about our genetic constitutions and ensure the most biologically appropriate mate. Or more likely, being the smart and adventurous species that we are, we'll experiment with nature and splash on a dab of someone different each night—and find out exactly what the nose knows.

Are You Ready for a New Sensation?

Kathryn S. Brown

The flimsy strip of golden film lying on John Wyatt's desk looks more like a candy wrapper than something you'd willingly put in your eye. Blow on it, and the 10-centimeter foil curls like cellophane. Rub it, and the shiny film squeaks faintly between your fingers. In fact, you have to peer rather closely to spot an unusual patchwork of squiggles: 100 electrodes, carefully arranged to jump-start cells in a damaged retina and, Wyatt hopes, allow the blind to see.

The film is part of a prototype retinal implant. For the past decade, Wyatt—an engineer at the Massachusetts Institute of Technology—and his colleagues have devoted a fifth-floor laboratory and countless hours to this tiny device. At first, even Wyatt doubted the project could succeed. The retina, he says, is more fragile than a wet Kleenex: it's a quarter of a millimeter thin and prone to tearing. In about 10 million Americans—those with disorders called retinitis pigmentosa and macular degeneration—the delicate rod and cone cells lining the retina's farthest edges die, although ganglion cells closer to the lens in the center survive. In 1988 Harvard Medical School neuro-ophthalmologist Joseph Rizzo asked Wyatt two key questions: Could scientists use electricity to jolt these leftover ganglion cells and force them to perceive images? Could they, in effect, engineer an electronic retina?

Try as he might, Wyatt couldn't think of a reason why the approach wouldn't work. Today Wyatt and Rizzo have tested their retinal implant on three patients. The most recent, a woman who participated in studies in spring 1999, reported seeing a four-dot design that perfectly matched the electrode stimulation to her retina. "Those were our best results yet," Wyatt remarks.

Despite these early returns, however, a practical working implant is still years away. Wyatt likes to call the project a "classic case of science: ten seconds of brilliance followed by ten years of dogged work."

When it comes to improving our senses, researchers have some truly brilliant ideas. In the coming years, if lab bench dreams become reality, we will see even when our eyes are damaged, hear even when our ears grow old, smell a whole new repertoire of scents and taste a much sweeter world. True, the goals are high and the technical hurdles steep. But the basic science is coming together today, as the worlds of engineering and biology blend. "Really, we are limited [only] by our imagination," claims Richard J. H. Smith, a molecular geneticist at the University of Iowa.

Smith's imagination travels to the recesses of the inner ear and the pea-size cochlea that holds some 16,000 noise-detecting cells, each of which is equipped with several hairlike projections that have earned them the name "hair cells." This precious stock of cells is a gift at birth: they never multiply, but they do die. Loud noise, disease and just plain aging damage hair cells, muffling one's ability to hear sounds that once seemed crystal clear.

Sensing the Future

Today people with poor hearing have two choices: a cochlear implant or an old-fashioned hearing aid. A cochlear implant is a surgically implanted set of electrodes that stimulates inner-ear cells, whereas a hearing aid is essentially a removable microphone and receiver. But researchers say these technologies—which basically turn up life's volume—are like using a sledgehammer to set a watch. In the future, scientists hope to coax the inner ear gently into repairing itself—or, better yet, to protect hair cells from damage in the first place.

Regenerating damaged or destroyed hair cells has gone from a science-fiction dream to a realistic hope. "Fifteen years ago if I'd applied for a grant to study hair cell regeneration, I'd have been laughed out of town," says Edwin W. Rubel of the University of Washington. "Now there are labs all over the world working on it."

One of the most promising approaches is to find genes that make hair cells grow and then pump them, via gene therapy, into a patient's ear. This may not be as hard as it sounds. Smith and other investigators have already discovered more than 25 specific gene sequences that are involved in hearing loss or deafness, and the search has just begun. By starting with easy-to-spot genetic mutations that cause extreme, inherited troubles, such as the progressive hearing loss that sometimes strikes college-age adults, researchers hope to find the genes that might also cause more widespread, age-related hearing loss.

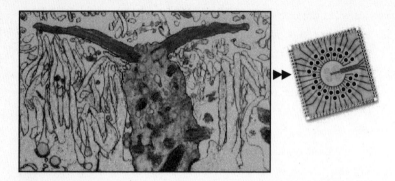

If someone tells you to wake up and smell the coffee, he or she might want you to use one of these. This orange blob is one of the thousands of olfactory receptors that make up the olfactory epithelium, a patch of mucous membrane way up in the nose that helps you sniff whether your milk has turned (among other things). Although the human nose isn't the best in the animal kingdom, researchers have mimicked it with a "nose on a chip" (right) that can be used by companies to monitor food quality. One day researchers might adapt the technology to develop an implant for people who have lost their sense of smell.

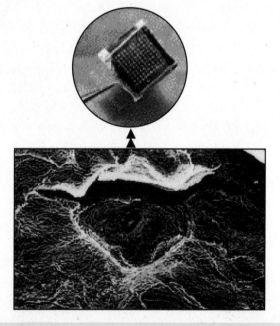

No, this isn't a close-up of one of those nubbly things on the surface of your tongue. Those are papillae; this is the opening of a taste bud. Hundreds of these barrel-shaped structures (seen here from above) are embedded in some types of papillae. When flavors enter the tiny pore in the center, they bind to and react with molecules called receptors on the surface of each of the taste cells, which make up the staves of the barrel. Scientists aren't producing an implantable artificial tongue just yet, but they have designed an electronic tongue, or e-tongue (top), that could be used to "taste" the quality of wine or the purity of water.

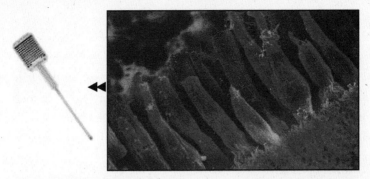

The rods and cones that make up the retina—the inside lining of the back of the eye—got their names for a reason that's obvious from this photograph. The rods are most important for black-and-white vision in dim light; the cones provide color vision and high visual acuity in bright light. But in people with diseases such as retinitis pigmentosa and macular degeneration, these sight cells start to die off, robbing the individuals of their vision. Bioengineers have now designed a retinal implant (above left) that could restore vision by allowing so-called ganglion cells, which are usually left intact in such diseases, to send electrical signals to the brain to register visual stimuli. The device is now being tested in people.

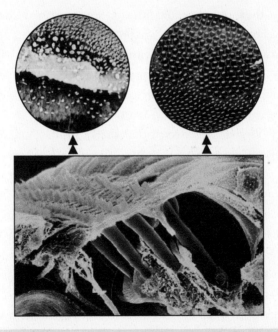

This detail from the cochlea, a tiny snail-shaped structure in the inner ear, reveals rows of sensory cells called hair cells. Each cell's minuscule projections register sounds and pass the information on to nerves that notify the brain. Exposure to loud noises and some drugs can destroy hair cells, causing hearing loss. Biologists are now trying to get damaged hair cells to regenerate. They've had some success with chicks: the electron micrographs above show hair cells disrupted by loud sounds (left) that have grown back 10 days later (right).

Other scientists are hunting for genes that are basic to hair cell development. In June 1999 geneticists at the Howard Hughes Medical Institute at Baylor College of Medicine, led by Huda Zoghbi, reported identifying a gene, named *Math1*, that is considered critical for the growth of hair cells in the inner ear. (*Math1* stands for m*ouse* at*onal* homoglog-*1*.) In their experiments, embryonic mice lacking *Math1* failed to develop hair cells at all. Adding extra copies of the human equivalent of *Math1* might trigger human hair cells to start growing again.

Once scientists know the correct genes to add, therapy becomes a matter of technique. Fortunately, Smith points out, the inner ear has two openings—the so-called round and oval windows—that doctors can use to shuttle genes into cells there. As with all gene therapy, scientists would have to find the right vectors—usually viruses engineered to carry an extra genetic payload—to get genes into specific cells. In some cases, physicians might bypass the faulty gene and instead simply repair the damage by, say, altering the chemical makeup of the fluid in the inner ear. "Depending on what we learn about hearing and genetics, we can come up with all kinds of creative ways to limit hearing loss or prevent it altogether," Smith predicts.

Some solutions might come from other animals. In 1974, during his first years of graduate school, Jeffrey T. Corwin, now a neuroscientist at the University of Virginia, discovered that sharks produce hundreds of thousands of hair cells throughout their lives. Corwin asked how—and whether human ears could be stimulated to do it, too. These questions still drive his research today.

Scientists now know that animals as diverse as zebrafish and chickens experience hair cell regeneration when their ears are damaged. By studying this faculty, investigators plan to pinpoint the key molecules involved, such as growth factors, and then design drugs based on the compounds. Even the runaway cell growth of cancer offers lessons in launching cell proliferation. If scientists learn how cancer nudges resting cells to suddenly start growing, they might also learn how to prompt hair cells to divide.

One day researchers could prevent hair cells from dying at all. With the right drug, predicts University of Virginia biomedical engineer Jonathan H. Spindel, it could be as simple as putting a few drops into someone's ear.

Some studies suggest that nerve cells in the cochlea will grow toward certain growth factors. If that is true, a modified cochlear implant might slowly release growth factors into the ear, luring nerve cells to multiply toward stimulating electrodes that would keep them growing and healthy.

Peering into the future, in fact, investigators toy with the idea of dispensing with hair cells altogether and instead implanting an array of electrodes directly into the brain's crevices or onto its surface, where the electrodes would spark the perception of hearing. This approach, Corwin notes, is rife with questions—among them, exactly where to put the electrodes and how to avoid damaging the brain. But biocompatible materials and compact computers keep improving. At this rate, he forecasts, "areas of opportunity that once were the exclusive domain of science-fiction authors may come into areas of medical practice."

An Artificial Nose?

For scientists who study smell, the world of nonfiction still holds many questions. Why can the scent of the family attic—or a stranger's perfume— prompt intense memories? How does your brain recognize a scent even before you can name it? And here's one that John S. Kauer really wants to answer: Why can't his wife smell the scent of the freesia flower?

Kauer, a neuroscientists at Tufts University, has been studying the olfactory system for 20 years, and he's still intrigued by anosmia, an absent or impaired sense of smell. Some people, like Kauer's wife, can't detect particular scents; others can barely smell anything at all. In fact, Kauer suggests, all human snouts could be missing out. "There is a world of [scent] molecules out there," he observes. "Just as there are animals that can see into the ultraviolet light or the infrared spectrum, there's likely a lot of odors we cannot smell."

Over the past few years Kauer and other scientists have been building "electronic noses": devices designed to sniff things we can't or might not want to, like land mines or spoiled food. Hewlett-Packard and Cyrano Sciences, a company based in Pasadena, California, for example, have designed an e-nose to help other companies monitor the quality of food and consumer products.

So far the e-noses only mimic human olfaction—and crudely at that, because each has just a few dozen sensors, compared with the millions of olfactory receptors in the human nose. But some scientists think that in the years to come, all this tinkering just might work in the other direction. "In a *Star Trek* kind of vision, you could imagine an artificial device that would allow you to recognize new scents in your environment," Kauer speculates. And just maybe, he posits, the device might live in a logical place: the lining of your own nose.

No matter how you engineer it, a stronger sniffer could improve life. Older adults whose sense of smell has gradually faded over the years often eat poorly, a reflection of the fact that most of food's flavor is really smell. According to the National Institute of Deafness and Other Communication Disorders in Bethesda, Maryland, more than 200,000 people in the U.S. visit a doctor for a smell or taste problem each year. And some of us might just want to enjoy the roses a bit more.

If Paul A. Grayson has his way, we'll soon get the chance. Grayson is president of an electric company called Ambryx in San Diego. Ambryx's goal is to turn today's molecular biology into a whole new field of products that pack a sensory punch. "What's missing from the twenty-first-century sensory experience?" Grayson asks. "The ability to enhance the sensory environment."

Run by a team of neuroscientists, corporate directors and even a cookbook author, Ambryx plans to bring sensory biochemistry to drug development and agricultural biotechnology, among other fields. One example might be perfume that's concocted according to a person's genetic profile. For example, a woman who can't smell musk—a common substance in perfumes—might prefer an undiluted jasmine scent. With DNA chip technology, companies could design a range of perfumes based on someone's unique olfactory receptor genes, says Peter Mombaerts, a neuroscientist at the Rockefeller University. In the summer of 1997 Ambryx announced a deal to use olfactory receptor genes discovered by Mombaerts's lab to look into such products.

Yummy Science

It seems only natural, perhaps, that Ambryx also wants to dabble in taste. In February a team led by Nicholas J. P. Ryba of the National Institute of Dental and Craniofacial Research in Bethesda, Maryland, and Charles Zuker of the University of California at San Diego reported a molecular coup: the discovery of two sentinel-like molecules on the surface of the tongue's taste cells that sense sweet and bitter flavors. Within two months of the announcement, Ambryx had licensed rights to conduct studies of the receptors.

Researchers currently know little about the molecules, which were dubbed TR1 and TR2, but they could hold the key to a new wave of medications that lack a bitter taste or of foods with a special sweetness. Ryba and his colleagues are now inactivating the genes that encode the two receptors in mice that will then be tempted with a smorgasbord of sweet and bitter treats to help confirm the receptors' flavorful roles. He says his lab will next begin hunting for receptors that sense salty and sour flavors.

Our sense of taste endures lifelong, Ryba says, so high-tech tongue implants aren't likely in the near future. But at least one research group has engineered a new spin on taste: the electronic tongue. Like the e-nose, the e-tongue takes a cue from human biology, using chemical sensors as artificial taste buds to sample less than appealing—or downright dangerous—fluids, such as blood or urine.

Ever since chemist John T. McDevitt and his colleagues at the University of Texas at Austin created the e-tongue in 1998, they have been peppered with ideas for using the device as diverse as wine tasting and virus assays. A Japanese travel agency even called to ask whether McDevitt could design an e-tongue to test the water in a foreign country to determine if it is safe for travelers to drink.

Could all this lead full circle to offer new ways to manipulate human taste? "That's an important direction for the science that we'd like to explore in the future," McDevitt comments. "But at this point, I just don't know."

One thing is certain. No matter what the goal, every lab that is blending electronics and biology—whether it's in the ear, on the brain, inside the nose or lining the eye—must figure out how to make human and machine communicate. M.I.T.'s Wyatt quips that the retina, which is sensitive to even the slightest pressure, doesn't welcome a brick of a microchip any more than you'd like being caressed by a bulldozer. The challenge is to stimulate sensory areas such as the retina *gently*.

Wyatt and Rizzo's retinal implant would do just that. The film, which slips inside a tiny incision made in the retina during a surgical procedure, has three thin layers: a 12-photodiode array to perceive light changes; a gold-colored strip with 100 electrodes to fire up retinal cells; and a stimulator chip that helps to direct current to the electrodes.

In the future, a patient who has received an implant will wear special glasses equipped with a miniature camera that captures images. The glasses will sport a small laser that receives the camera's pictures and converts the visual information into electrical signals that travel to the implant. The implant, in turn, will activate the retina's ganglion cells to pick up the sensation of the image coming in and convey it to the brain, where it will be perceived as vision.

If it sounds complicated, Wyatt comments dryly, that's because it is. Nevertheless, he and his colleagues have been slowly perfecting the technique over a series of experiments—lengthening the duration of the current pulses and fine-tuning the microelectrode arrays. One looming question is how the retinal implant will work over time. So far the researchers have performed only afternoonlong experiments, after which the microelectric array is removed.

Since Wyatt and Rizzo's work began, two other groups in the U.S. have taken up the cause. One is Optobionics, a start-up company headed by Wheaton, Illinois, ophthalmologist Alan Y. Chow. Optobionics is now testing its implant, which is named the artificial silicon retina, in rabbits. The Optobionics device is a subretinal implant, meaning it's surgically implanted beneath the retina. It is different from the M.I.T. group's retinal implant in that it connects to the back side—the photoreceptor side—of the retina rather than to ganglion cells. The second team, a group of scientists at Johns Hopkins University and at North Carolina State University, is pursuing a retinal implant similar to Wyatt and Rizzo's. The device is promising, although researchers must still demonstrate its long-term biocompatibility with the tissues of the eye, says Wentai Liu of N.C.S.U.

Although it is unusual today, an artificial retina could fit quite comfortably into the bionic body of tomorrow. Eventually, Liu predicts, investigators might create miniature computer chips that can be integrated fully into the body, allowing someone to recover from any injury with the help of internal electronic signals. "That's the next century," he says. "Right now we'll be very excited if we can just help people recover their sight."

Index

Photo Credits: Page 17: New Bone Grows to fill a Space © 2001 Laurie Grace and Keith Kasnot. Page 59: Health Care Expenditures © 2001 Roger Doyle. Page 73: Implant © 2001 Robert Osti. Page 77: Promising Liver Support Approach © 2001 Robert Osti. Page 97: Two Genome-Sequencing Strategies © 2001 Laurie Grace. Page 107: Function of Essential Genes © 2001 Victoria Roberts, Arcady R. Mushegian, and Eugene V. Koonin. Page 119: Delivery of Genes © 2001 Slim Films. Page 123: Naturally Occurring Virus © 2001 Tomo Narashima. Page 173: Dumb Mouse/Smart Mouse © 2001 Terese Winslow. Page 185: Split and Mix Synthesis © 2001 Jared Schneidman Design. Page 197: Male-Female Fertilized Eggs © 2001 Dmitry Karsny and Peter Lawrence. Page 208: ChartShowing Calculations per Second © 2001 Laurie Grace. Page 224: Computer Chip and Biological Sensory © 2001 P. Motta SPL/Photo Researches Inc., Cyrano Sciences, John T. McDevitt University of Texas at Austin. Pages 25: Computer Chiop and Biological Sensory © 2001 P. Motta SPL/ Photo Researches Inc., Cyrano Sciences, John T. McDevitt University of Texas at Austin.